Lecture Notes in Physics

Bisher erschienen/Already published

Vol. 1: J. C. Erdmann. Wärmeleitung in Kristallen, theoretische Grundlagen und fortgeschrittene experimentelle Methoden. II, 283 Seiten. 1969.

Vol. 2: K. Hepp, Théorie de la renormalisation. III, 215 pages. 1969.

Vol. 3: A. Martin, Scattering Theory: Unitarity, Analyticity and Crossing. IV, 125 pages. 1969.

Vol. 4: G. Ludwig, Deutung des Begriffs „physikalische Theorie" und axiomatische Grundlegung der Hilbertraumstruktur der Quantenmechanik durch Hauptsätze des Messens. 1970. Vergriffen.

Vol. 5: Schaaf, The Reduction of the Product of Two Irreducible Unitary Representations of the Proper Orthochronous Quantummechanical Poincare Group. IV, 120 pages. 1970.

Vol. 6: Group Representations in Mathematics and Physics. Edited by V. Bargmann. V, 340 pages. 1970.

Vol. 7: R. Balescu, J. L. Lebowitz, I. Prigogine, P. Résibois, Z. W. Salsburg, Lectures in Statistical Physics. V, 181 pages. 1971.

Vol. 8: Proceedings of the Second International Conference on Numerical Methods in Fluid Dynamics. Edited by M. Holt. 1971. Out of print.

Vol. 9: D. W. Robinson, The Thermodynamic Pressure in Quantum Statistical Mechanics. V, 115 pages. 1971.

Vol. 10: J. M. Stewart, Non-Equilibrium-Relativistic Kinetic Theory. III, 113 pages. 1971.

Vol. 11: O. Steinmann, Pertubation Expansions in Axiomatic Field Theory. III, 126 pages. 1976.

Vol. 12: Statistical Models and Turbulence. Edited by C. Van Atta and M. Rosenblatt. Reprint of the First Edition. VIII, 492 pages. 1975.

Vol. 13: M. Ryan, Hamiltonian Cosmology. VII, 169 pages. 1972.

Vol. 14: Methods of Local and Global Differential Geometry in General Relativity. Edited by D. Farnsworth, J. Fink, J. Porter, and A. Thompson. V, 188 pages.

Vol. 15: M. Fierz, Vorlesungen zur Entwicklungsgeschichte der Mechanik. V, 97 Seiten. 1972.

Vol. 16: H.-O. Georgii, Phasenübergang 1. Art bei Gittergasmodellen. IX, 167 Seiten. 1972.

Vol. 17: Strong Interaction Physics. Edited by W. Rühl and A. Vancura. V, 405 pages. 1973.

Vol. 18: Proceedings of the Third International Conference on Numerical Methods in Fluid Mechanics, Vol. I. Edited by H. Cabannes and R. Temam. VII, 186 pages. 1973.

Vol. 19: Proceedings of the Third International Conference on Numerical Methods in Fluid Mechanics, Vol. II. Edited by H. Cabannes and R. Temam. VII, 275 pages. 1973.

Vol. 20: Statistical Mechanics and Mathematical Problems. Edited by A. Lenard. VIII, 247 pages. 1973.

Vol. 21: Optimization and Stability Problems in Continuum Mechanics. Edited by P. K. C. Wang. V, 94 pages. 1973.

Vol. 22: Proceedings of the Europhysics Study Conference on Intermediate Processes in Nuclear Reactions. Edited by N. Cindro, P. Kulišic and Th. Mayer-Kuckuk. XIV, 329 pages. 1973.

Vol. 23: Nuclear Structure Physics. Proceedings 1973. Edited by U. Smilansky, I. Talmi, and H. A. Weidenmüller. XII, 296 pages. 1973.

Vol. 24: R. F. Snipes, Statistical Mechanical Theory of the Electrolytic Transport of Nonelectrolytes. V, 210 pages. 1973.

Vol. 25: Constructive Quantum Field Theory. The 1973 "Ettore Majorana" International School of Mathematical Physics. Edited by G. Velo and A. Wightman. III, 331 pages. 1973.

Vol. 26: A. Hubert, Theorie der Domänenwände in geordneten Medien. XII, 377 Seiten. 1974.

Vol. 27: R. K. Zeytounian, Notes sur les Ecoulements Rotationnels de Fluides Parfaits. XIII, 407 pages. 1974.

Vol. 28: Lectures in Statistical Physics. Edited by W. C. Schieve and J. S. Turner. V, 342 pages. 1974.

Vol. 29: Foundations of Quantum Mechanics and Ordered Linear Spaces. Advanced Study Institute, Marburg 1973. Edited by A. Hartkämper and H. Neumann. VI, 355 pages. 1974.

Vol. 30: Polarization Nuclear Physics. Proceedings 1973. Edited by D. Fick. IX, 292 pages. 1974.

Vol. 31: Transport Phenomena. Sitges International Schools of Statistical Mechanics, June 1974. Edited by G. Kirczenow and J. Marro. XIV, 517 pages. 1974.

Vol. 32: Particles, Quantum Fields and Statistical Mechanics. Proceedings 1973. Edited by M. Alexanian and A. Zepeda. V, 132 pages. 1975.

Vol. 33: Classical and Quantum Mechanical Aspects of Heavy Ion Collisions. Proceedings 1974. Edited by H. L. Harney, P. Braun-Munzinger, and C. K. Gelbke. VII, 311 pages. 1975.

Vol. 34: One-Dimensional Conductors GPS Summer School Proceedings, 1974. Edited by H. G. Schuster. VII, 371 pages. 1975.

Vol. 35: Proceedings of the Fourth International Conference on Numerical Methods in Fluid Dynamics, 1974. Edited by R. D. Richtmyer. V, 457 pages. 1975.

Vol. 36: R. Gatignol, Théorie Cinétique des Gaz à Répartition Discrète de Vitesses. II, 219 pages. 1975.

Vol. 37: Trends in Elementary Particle Theory. Proceedings 1974. Edited by H. Rollnik and K. Dietz. V, 472 pages. 1975.

Vol. 38: Dynamical Systems, Theory and Applications. Proceedings 1974. Edited by J. Moser. VI, 624 pages. 1975.

Vol. 39: International Symposium on Mathematical Problems in Theoretical Physics. Proceedings 1975. Edited by H. Araki. XII, 562 pages. 1975.

Vol. 40: Effective Interactions and Operators in Nuclei. Proceedings 1975. Edited by B. R. Barrett. XII, 339 pages. 1975.

Vol. 41: Progress in Numerical Fluid Dynamics. Proceedings 1974. Edited by H. J. Wirz. V, 471 pages. 1975.

Vol. 42: H II Regions and Related Topics. Proceedings 1975. Edited by D. Downes and T. L. Wilson. XII, 488 pages. 1975.

Vol. 43: Laser Spectroscopy. Proceedings 1975. Edited by S. Haroche, J. C. Pebay-Peyroula, T. W. Hänsch, and S. E. Harris. X, 466 pages. 1975.

Lecture Notes in Physics

Edited by J. Ehlers, München, K. Hepp, Zürich,
R. Kippenhahn, München, H. A. Weidenmüller, Heidelberg,
and J. Zittartz, Köln
Managing Editor: W. Beiglböck, Heidelberg

67

W. Drechsler
M. E. Mayer

Fiber Bundle Techniques in Gauge Theories

Lectures in Mathematical Physics at
the University of Texas at Austin
Edited by A. Böhm and J. D. Dollard

Springer-Verlag
Berlin Heidelberg GmbH 1977

Authors
Dr. Wolfgang Drechsler
Max-Planck-Institut
für Physik und Astrophysik
Föhringer Ring 6
8000 München 40
BRD

Dr. M. E. Mayer
Department of Physics
University of California
Irvine, CA 92717
USA

Editors
Prof. A. Böhm
Physics Department
University of Texas
Austin, TX 78712
USA

Prof. J. D. Dollard
Mathematics Department
University of Texas
Austin, TX 78712
USA

Library of Congress Cataloging in Publication Data

Drechsler, Wolfgang, 1934-
Fibre bundle techniques in gauge theories.

(Lecture notes in physics ; 67)
Bibliography: p.
Includes index.
1. Gauge fields (Physics)--Addresses, essays, lectures. 2. Fiber bundles (Mathematics)--Addresses, essays, lectures. I. Mayer, Meinhard Edwin, 1929- joint author. II. Title. III. Series.
QC793.3.F5D73 530.1'43 77-23936

ISBN 978-3-540-08350-4 ISBN 978-3-540-37289-9 (eBook)
DOI 10.1007/978-3-540-37289-9

Originally published by Springer-Verlag Berlin Heidelberg New York in 1977

2153/3140-543210

Preface

The two contributions in this volume originated in lecture series given in the mathematical physics program at the University of Texas at Austin in 1976 and 1977. The purpose of this program is to help establish communication between mathematicians and physicists and to inform graduate students of both departments about recent developments in mathematics and physics that may be of use to them. It is hoped that this program will provide inspiration for both groups, by introducing new mathematical structures which may prove useful to physicists, and by exposing physical problems which may require the creation of further mathematical structures. All lectures were directed at "non-specialists", so this volume should be suitable for a general audience.

Both contributions in this volume are concerned with fibre bundles, and in particular their application to the study of gauge groups. As the lectures were prepared and delivered independently, there is some overlap between them. However, there is a clear distinction between the two contributions. The lectures of M. E. Mayer emphasize the mathematics, while the lectures of W. Drechsler appeal more to the intuition of the physicist. Although the reformulation of conventional gauge theories in the language of modern differential geometry would be an interesting exercise, the real hope is that the effort required will be amply repaid by a deeper insight into existing theories and by the ability to apply this language to other problems. Fibre bundles may eventually give us a means of describing hadrons as extended yet elementary objects appearing as domains of strong curvature in a bundle constructed over space-time, and this would more than justify any effort required to learn the new language. Furthermore, fibre bundles provide a convenient framework for discussing the concepts of relativity and invariance. After having come to understand the concept of the fibre bundle, the physicst will perhaps be surprised to find how often this concept can be naturally employed in physical situations. The mathematician who reads this volume may be gratified and inspired to see examples of how these abstract mathematical structures find application in physics.

This first volume of lecture notes from the Texas Mathematical Physics lecture series is dedicated to Paul Olum, who was Dean of Natural Sciences at the inception of the series. Dean Olum not only made these lectures possible by his enthusiastic support, but was also the first to tell us that physicists should know about fibre bundles and to express to us the belief that they could play an important role in the formulation of physical theory.

A. Bohm
J. D. Dollard
(Editors)

(Detailed tables of contents for each set of notes can be found on pages 5 and 146, respectively)

TABLE OF CONTENTS

PART I

PART II

PART ONE

INTRODUCTION TO THE FIBER-BUNDLE APPROACH TO GAUGE THEORIES

(Lectures delivered at the University of Texas
at Austin in May 1976, at the University
of California Irvine in 1975/76, and at
the Aspen Center for Physics, June 1976)

"Jede Wissenschaft ist, unter andrem,
ein Ordnen, ein Vereinfachen, ein
Verdaulichmachen des Unverdaulichen
für den Geist".

Hermann Hesse

(Every science is, among other things,
the ordering, the simplifying, the
making digestible what is undigestible
for the spirit.)

PREFACE

These notes are the outgrowth of a seminar given in the academic year 1975/76 at the University of California, Irvine, a series of lectures given in May 1976 at the Center for Particle Theory of the University of Texas at Austin, and a lecture on characteristic classes and pseudoparticles (instantons), given in June 1976 at the Aspen Center for Physics.

In writing up the notes, I have attempted to expand the sections dealing with differential-geometric problems, but without going into detailed proofs. The main aim has been to make these developments accessible to physicists, and to whet their appetites for more detailed reading. Unfortunately, time and space limitations have not allowed me to develop in more detail some of the topics in which I am most interested, such as quantization schemes for gauge fields in terms of quantized differential forms, the role of characteristic classes in quantization (in addition to their role in classifications of classical solutions). I hope to be able to cover these topics in another lecture note volume, based on lectures which I hope to deliver at the ETH during the winter semester 1977/78.

The material covered in these notes should be clear from the table of contents and the index. I have not attempted to provide the reader with an exhaustive bibliography; in particular, I have not tried to list the large number of papers on applications of gauge theory, which have appeared in the major physics journals in the past few years. The big review articles quoted in the list of references should be consulted for these. The bibliography is geared to a large degree to the historical survey given in the Introduction, which is highly subjective and incomplete. I would like to apologize in advance to any author whose work has not been quoted, or inadequately quoted. I am fully aware of possible omissions, but time-pressure has prevented me from undertaking a systematic search of the published and unpublished literature.

I am indebted to many people for making this set of notes possible. First of all, I would like to thank Professor Konrad Bleuler, whose invitations to the Bonn Symposia in 1973 and 1975 rekindled my interest in gauge theories and fiber bundles. I would like to thank Raoul Bott for some very illuminating discussions on characteristic classes in 1976, George Sudarshan and Arno Böhm for inviting me to lecture on this topic in Austin, my colleagues and students at Irvine for listening patiently to my lectures and making things clearer with their questions. I am grateful to Howard Abrams, Thomas Erber, Mike Kovacich and several others for pointing out some misprints and syntactic errors in the first three chapters of these notes (i am solely responsible for the surviving errors in the text, typographical or otherwise). Professor I. M. Singer told me about the results described in Section 6.2.1 and let me have a preprint of ref. [65], for which I am greatly indebted. I wish to thank my daughter Elma for typing part of the text. Finally, I wish to express my appreciation to Professors K. Hepp and W. Beiglböck for encouraging me to write up this material for the Lecture Notes in Physics.

Corona del Mar, California

May, 1977

PART I

TABLE OF CONTENTS

0. INTRODUCTION

If there has been one unifying feature of elementary particle theories during the last decade it was the discovery that gauge theories are probably the best candidates for a genuine theory of elementary particles, since they have allowed, on the one hand, to unify weak and electromagnetic interactions, and on the other hand, they hold out some hope for understanding the quark-gluon picture of strong interactions, quark confinement, etc. Combine this with the insight that the theory of gravitation, in its Einsteinian form, is ultimately also a gauge theory and that the latter has been with us for over 60 years, and you will be surprised that physicists have not started paying attention earlier to the beautiful geometric concepts on which gauge theory is based.

This is all the more surprising, as the term gauge invariance and the basic idea of "local gauge transformations" has been invented by Hermann Weyl in 1918, and extended in 1929 [61] to a theory of the electromagnetic field in interaction with charged particle fields. Although the gauge ambiguity of the electromagnetic field potentials had been known for some time, one must consider Emmy Noether's paper [40] on invariant variational principles as the precursor of present-day gauge theories. The early formulations of quantum electrodynamics [16],Pauli, recognized the difficulty in quantizing the electromagnetic field due to the contradictions between gauge invariance and Lorentz invariance requirements for the potentials, difiiculties which have led in the 50's to the development of the Bleuler-Gupta quantization [6, 22], and which have been properly resolved only quite recently.[51]

Although the idea of a local gauge transformation (the transformation of a field under an internal symmetry group with parameters depending on the point) can and is formulated usually in a classical context, its full impact is not felt until one considers quantum theories. This is probably why Weyl's gauge principle was almost forgotten for 20 years, until Schwinger[48] treated the electromagnetic field as a consequence of local U(1)-gauge-invariance for the quantized Dirac field. This led directly to Maxwell's equations and to their quantization in the Coulomb gauge.

Soon thereafter, C. N. Yang and R. L. Mills [63] extended this idea to a field theory of isospin-invariant particles (nucleons or pions with SU(2)-invariant couplings) and were thus the first to discover the existence of a triplet of vector fields which now bear their names,and of the generalization of the Maxwell equations to nonabelian gauge groups, which are now known as the Yang-Mills equations. A reading of the Yang-Mills paper shows that the geometric meaning of the gauge potentials must have been clear to the authors, since they use the gauge-covariant derivative and the curvature form of the connection, and indeed, the basic equations in that paper will coincide with the ones derived from a more geometric approach in Section 2.3.

The Yang-Mills approach was generalized in 1955-56 to more general gauge groups (arbitrary compact Lie groups) independently by Utiyama[57] and by the author in his dissertation[36], Mayer. It was recognized that Gell-Mann's principle of minimal coupling requires replacing the derivatives in Lagrangians or field equations by gauge-covariant derivatives and Utiyama stressed the geometric character of the gauge principle more than any other author at that time. He actually pointed out that the Christoffel symbols of general relativity can be considered as gauge potentials if one subjects the Lorentz group itself to a local gauge transformation, and the Einstein equations follow as Yang-Mills equations (or Bianchi identities) for the curvature tensor. This point of view was further investigated by Thirring[52], Kibble[29], and others, and is still being discussed in the literature now.

Gauge theory underwent a rapid development in the period 1957 - 1961 (here is an incomplete sampling of papers which come to mind:

[2, 20, 29, 41, 46, 52])and played an important role in the discovery of the SU(3)-symmetry by Neeman and Gell-Mann. A large number of papers was devoted to the quantization of gauge theories and to finding solutions of Yang-Mills equations [12, 15, 17, 31,58], but it was not until 1967/68 that Weinberg [59] and Salam [47] discovered the unified theory of weak and electromagnetic interactions, which they suspected of being renormalizable, and 't Hooft [53] showed this to be indeed the case (cf. the reviews [1, 60]). This model was made possible through the discovery of the so-called Higgs mechanism for symmetry breaking [9, 23, 25, 30], which allowed the vector fields to acquire a mass and gets rid of unwanted Goldstone bosons (cf. the review [5]). I will not attempt to list the deluge of post-1970 papers on gauge theories, and just refer the reader to almost any issue of the major physics journals.

Relatively few people paid attention (until recently) to the geometric and topological aspects of gauge theories. Early attempts in this direction were made by Lubkin [33], who pointed out the fiber-bundle structure of a gauge theory and Loos [31], who emphasized the role of the "internal holonomy groups"; the bundle concept was also emphasized by Robert Hermann in his numerous publications.

Although I had been interested in gauge theories since 1955 and had looked into the fiber-bundle aspects of gauge fields as ealy as 1965 [37] in response to a question asked by George Sudarshan, my interest in the subject was reactivated by a lecture of A. Trautman [47] at a Symposium in Bonn in 1973. It resulted in a number of contributions on which these notes are based [37],and a renewed interest in the quantization of gauge field theories, when one considers the gauge potentials as connection forms in a principal fibration. The use of connections was advocated by Faddeev[14], Yang [62,64], and others.[18,27]

In 1974 Polyakov [43] and 't Hooft (cf. the lecture notes of S. Coleman [11] for a complete bibliography) discovered some solutions of the classical Yang-Mills equations (called respectively "hedgehog" or 't Hooft monopoles) which exhibit "topological quantum numbers", i. e., homotopy types, given by the mappings which these solutions realize from the gauge group into various spheres at infinity (depending on the dimensionality of the Euclidean space on which the solutions were found). These discoveries then led to a number of very exciting results, due to Polyakov and coworkers [4], 't Hooft [54], Jackiw and Rebbi [26] and others [10, 34, 35], which discuss properties of the gauge theory "vacuum" and classify these vacua according to characteristic classes of the appropriate bundles [38]. Cf. also [65 - 70].

Another line of development in gauge theories, which is very promising, but which unfortunately we did not have time to go into, are the so-called lattice-gauge theories, proposed by Wilson, and developed in [7, 39]. Recently, a number of rigorous results have been obtained in this direction [21, 42], and the role of cohomology theories and characteristic classes for lattice gauge theories is under active investigation, at least by the author.

These notes are intended as an introduction to some of the later topics mentioned in this introduction. It is assumed that the reader is not familiar with coordinate-free differential geometry, and the first chapter is devoted to an introduction of the basic concepts. Readers who are familiar with differential forms might want to start with Chapter 2, which discusses electromagnetism in a coordinate-free notation and gives a heuristic introduction to the gauge principle, in order to motivate the notion of principal fibration, vector bundles and connections. Chapter 3 discusses principal fibrations and associated vector bundles to the extent needed for the formulation of gauge field theories. Examples are relegated to Chapter 6. Chapter 4 deals with connections in principal and associated fibrations, where connections are defined first as covariant differentials of sections of a vector bundle, and only later as a Lie-algebra valued form. Chapter

5 deals with an introduction to characteristic classes, the treatment following the rather elementary approach of Bott and Chern. Some general facts are mentioned, but in order not to overburden the reader with mathematical facts, we have restricted ourselves to a minimum. Finally, Chapter 6 contains all the physical applications, in particular, a brief discussion of quantization schemes, as well as of topological quantum numbers and their relations to the Chern classes.

In line with the aim of these notes, we have tended to emphasize the mathematical concepts over the physical applications, and hope to return to the latter elsewhere. As already mentioned, the list of references is far from exhaustive, and is geared mainly to this introduction, and books are quoted by the name of the author only.

The notations used are close to the standard ones, although typographical convenience has induced me occasionally to replace Greek letters by Latin ones. Lack of time has prevented me from preparing an index, but the table of contents is detailed enough, and new concepts are underlined in the text.

1. MANIFOLDS AND DIFFERENTIAL FORMS

1.0. Introduction. In this chapter we review briefly the basic concepts of differential geometry, differential manifolds and differential forms. The main purpose here is to familiarize the physicist with coordinate-free notation and with some of the jargon of modern differential geometry. Most concepts will be briefly illustrated with examples familiar in tensor notation from mechanics or electromagnetic theory and designed to provide a "dictionary" between the two notations. As a rule proofs are omitted or only sketched and the reader is referred to the list of textbooks at the end for references (textbooks are quoted by the name of the author). The examples are to be treated as exercises and worked out. We deliberately avoid discussing Riemannian geometry, which can be found in the standard texts on general relativity (in coordinate-free notation,e.g., in Misner-Thorne-Wheeler and Hawking-Ellis). On the other hand, a certain familiarity with the traditional formulation of general relativity is assumed on the part of the reader.

1.1. (Smooth) Differential Manifolds and their Tangent Bundles.

Roughly speaking, a differential manifold is locally a finite-dimensional vector space, or can be thought of as the result of glueing together in a smooth manner an infinity of such vector spaces (we shall see later that this concept appears again). This is made more precise in the following string of definitions.

1.1.1. Definition. Let X be a set. A chart of X is a triple

$$c = (U, \varphi, E),$$

where U is a part of X, E is a Banach space (in particular $\mathbb{R}^n$) and φ is a bijection of U onto an open set of E. U is called the domain of the chart. If $E = \mathbb{R}^n$, n is called the dimension of the chart. We will consider only finite-dimensional charts in the sequel, although many results are valid for infinite-dimensional Banach spaces.

1.1.2. Two charts c, c' are compatible (or C^∞-compatible) if:

i) $\varphi(U \cap U')$ and $\varphi'(U \cap U')$ are open in E, respectively E';

ii) the mapping $\varphi \circ \varphi'^{-1}$ of $\varphi'(U \cap U')$ onto $\varphi(U \cap U')$ (and its inverse $\varphi' \circ \varphi^{-1}$) is C^∞ (i.e., infinitely differentiable; differentiability of functions from one vector space to another is best thought of in terms of coordinates; for a general definition cf., e.g., Dieudonné, ch.VIII).

Remark. All our definitions will be given assuming infinite differentiability. Functions and manifolds satisfying this condition are usually called smooth. Most definitions are easily modified to accomodate C^r-differentiability, with finite r, but the statements of many results become more involved, and care is required not to exceed the order of differentiability. Since in quantum field theory we deal with distributions, the C^∞ assumption represents no loss of generality, and where necessary (e.g., near sources of fields) we will remove the set where a function ceases to be smooth, possibly at the cost of ruining the connectivity of the manifold under consideration. This will be seen to have a certain usefulness.

1.1.3. Definition. An atlas of a set X is a collection of charts of X which are mutually compatible, and for which the domains have X as union (i. e., a collection of compatible charts which "covers" X).

1.1.4. Definition. A real differential manifold of class C^∞ , or simply manifold, is a set X equipped with a collection of equivalent atlases.

It can be shown that X is a topological Hausdorff space which is locally connected (i. e., consists of connected pieces). If X is connected (i. e., any two points in X can be joined by a piecewise smooth curve), then all charts have the same dimension, the dimension of X.

<u>1.1.5. Examples.</u> a) $\mathbb{R}^n$, the n-dimensional vector space of real n-uples $(x_1,\ldots,x_n)$ is a manifold with an atlas consisting of the one chart: $c = (\mathbb{R}^n, \mathrm{Id}, \mathbb{R}^n)$, where Id is the identity map.

b) $S^2 = \{(x, y, z) \in \mathbb{R}^3 : x^2 + y^2 + z^2 = 1\}$, the unit sphere in 3-space is a manifold with one atlas given by two charts obtained, e.g., by stereographic projection onto $\mathbb{R}^2$ once from the north pole and once from the south pole. Alternatively, two charts of an equivalent atlas are obtained by using two spherical coordinate systems with different polar axes.

c) Fig. 1 illustrates the definition of a two-dimensional manifold.

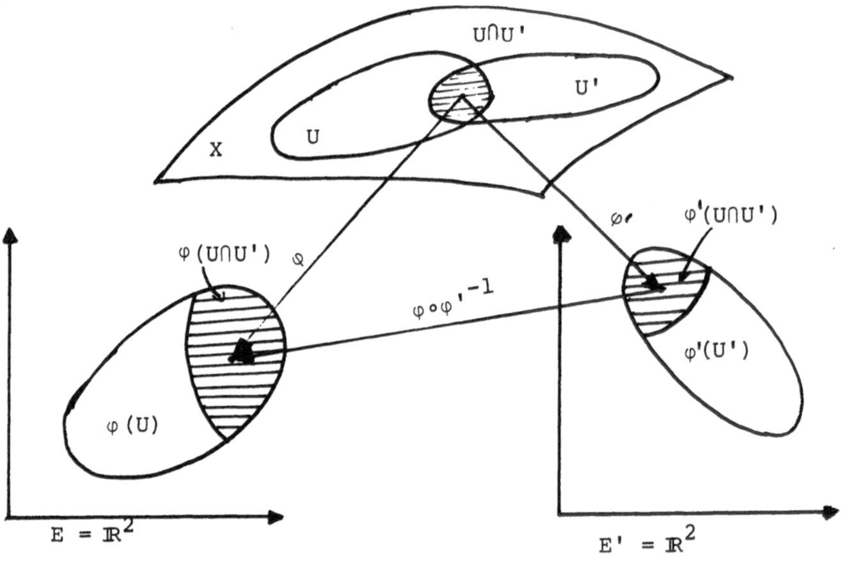

Fig. 1

d) The configuration space of a mechanical system is a manifold. In particular, the configuration space of the plane double pendulum is a torus $T^2 = S^1 \times S^1$ (use the angles as parameters).

e) <u>Embedded manifold</u>. A k-dimensional manifold M, embedded in $\mathbb{R}^n$ is defined by n - k functions $f_i : U \to \mathbb{R}$, where U is a subset of $\mathbb{R}^n$, as the set where $f_1 = 0,\ldots,f_{n-k} = 0$ and the vectors grad $f_1,\ldots,$ grad f_{n-k} are linearly independent.

f) A <u>Lie group</u> is a group which is also a manifold and such that the group operations (multiplication and inverse) are C^∞ functions.

g) The manifold SO(3) can be embedded in $\mathbb{R}^9$ (the space of all 3×3 matrices with real entries).

<u>1.1.6. Definition</u>. A mapping of a manifold X onto a manifold Y is a <u>diffeomorphism</u> if it is a bijection and both the mapping and its inverse are C^∞.

<u>1.1.7. Tangent space</u>. If the manifold M of dimension k is embedded in $\mathbb{R}^n$ it is clear that the k-dimensional tangent space at a point x, T_xM is the orthogonal complement to the normal vector space spanned by the gradients of Example 1.1.5, e. In order to obtain an intrinsic definition, we proceed as follows.

Consider two "curves" $x = f_1(t)$, $x = f_2(t)$ in the manifold M (i. e., smooth mappings of the segment $0 \leq t \leq 1$ into M). We say that the two curves are <u>tangent</u> at the point x if $f_1(a) = f_2(a) = x$ and if in some chart $c = (U, \varphi, \mathbb{R}^k)$ containing the point x they satisfy

$$\lim_{t\to a} \frac{\varphi\circ f_1(t) - \varphi\circ f_2(t)}{t - a} = 0.$$

(This statement is true in any chart if it is true in one chart.) The property of the two curves being tangent at x is an equivalence relation and the equivalence class of curves tangent to each other at x define a <u>tangent vector</u> to M at x. The set of all tangent vectors at x is denoted by T_xM. The above definition shows that there is a bijection $\theta_c\colon T_xM \to \mathbb{R}^k$ (mapping a neighborhood of the point a in [0, 1] into the gradient $D(\varphi\circ f)(a)$ we obtain the linear bijection θ_c in c).

<u>Examples.</u> a) The tangent plane to a sphere, torus, etc.

b) The tangent space to a Lie group G at the group identity e, T_eG, is the vector space which spans the Lie algebra g of G. We will identify g with T_eG, although the Lie algebra has an additional operation -- the bracket-- induced by the group composition.

Let $f:X \to Y$ be a morphism of the manifold X into the manifold Y (i. e., a C^∞ function which for each chart (V, ψ, F) of Y is such that the mapping of the open submanifold $f^{-1}(V)$ into the vector space F given by $\psi \circ f$ is C^∞). In other words in two charts (U, φ, E) of X and (V, ψ, F) of Y (E, F are the two vector spaces on which our manifolds are modeled, not necessarily of the same dimension) the function $\Phi = \psi \circ f \circ \varphi^{-1}$ is infinitely differentiable. The derivative of this function defines (in the given charts) the derivative of the morphism f, also called the tangent mapping, between the tangent spaces T_xX and $T_{f(x)}Y$ and denoted by

$$f_* = T_x f = \theta_{c'}^{-1} \circ D\Phi(\varphi(x)) \circ \theta_c \ . \tag{1.1}$$

It can be shown that this linear map between the tangent spaces does not depend on the charts c, c' chosen in the two manifolds and that it obeys the chain rule (g is a morphism of the manifold Y into Z)

$$T_x(g \circ f) = T_{f(x)} g \circ T_x f. \tag{1.2}$$

1.1.8. The tangent bundle. The union of all the tangent spaces of a manifold X

$$\bigcup_{x \in X} T_x X = TX$$

is called the tangent bundle of X and is denoted by TX (sometimes, in order to distinguish the tangent space better from the tangent bundle, the former will also be denoted by $T_x(X)$). If X is a differential manifold of dimension n, the tangent bundle TX is a manifold of dimension 2n. In a chart c of X a "point" of X is characterized by a vector in $\mathbb{R}^n$, or by its n coordinates $x_1,\ldots,x_n$. This produces a chart of TX, in which a "point" is characterized by 2n coordinates: the n coordinates $x_1,\ldots,x_n$ and the n components of the tangent vector $\xi \in T_xX$: $\xi_1,\ldots,\xi_n$. There is a C^∞-mapping from the tangent bundle TX to the manifold X which associates to the pair $\{x,\xi\}$ the point $x \in X$, denoted by π and called the projection of the bundle. The inverse image of x under the projection, $\pi^{-1}(x) = T_xX$ is called the fiber of the bundle and is obviously the tangent space at x to the manifold X. Locally, (i.e. in a chart $c = (U, \varphi, \mathbb{R}^n)$, the restriction of TX to U is diffeomorphic to the cartesian product $U \times \mathbb{R}^n$. This construction is the prototype for all bundles to be considered later and should be kept in mind as such. The manifold X is called the base space of the bundle and the tangent bundle can be thought of as a copy of $\mathbb{R}^n$ "sitting" over each point of X , which is itself an element of a copy of $\mathbb{R}^n$.

1.1.9. Example. The phase space of a mechanical system. Consider a mechanical system of N mass points moving in $\mathbb{R}^3$ subject to k constraints described by k real-valued smooth functions on $\mathbb{R}^{3N}$. The configuration space of this system with $n = 3N - k$ degrees of freedom is an n-dimensional differential manifold (embedded in $\mathbb{R}^{3N}$) and the motion of the system can be considered as a curve (mapping of the real line) in this manifold M. For simplicity let us assume that all particles have the same mass, which we set equal to one, so that momentum and velocity can be identified. Then a point on the trajectory together with its tangent vector can be identified with a point in the space of the tangent bundle TM, which thus becomes the phase space of the system. The Lagrangian L of such a mechanical system is a differentiable function $L: TM \to \mathbb{R}$. The manifold M of a dynamical system has an additional structure: the existence at each point of a positive definite quadratic form (making it into a Riemannian manifold), such that in each tangent space we have a quadratic form in the velocities, the kinetic energy $T = \frac{1}{2}\langle v, v\rangle$ (for details, cf., e. g., Abraham-Marsden or Arnol'd).

1.1.10. Vectors and tensors. A tangent vector to a manifold was defined above as an equivalence class of curves having a contact of first order. A vector field can be defined as the selection of a tangent vector at each point of the manifold, operation which is called forming a cross section (or section) of the tangent bundle, i. e., a section is a (smooth) map $s: M \to TM$ with the property that $\pi \circ s =$ identity, π being the projection of TM onto M. Alternatively, a vector field can be viewed as a directional derivative X, such that in a chart with local coordinates x^i we have for any differentiable function f:

$$Xf = \sum X^i \frac{\partial f}{\partial x^i}. \tag{1.3}$$

Linear forms on tangent vectors are called <u>1-forms</u> or <u>covectors</u> and will be denoted by lower-case greek letters:

$$\omega(X) = \langle \omega, X \rangle . \tag{1.4}$$

The space of all such linear forms at a point of M is called the <u>cotangent space</u> T_x^*M, and the totality of all these spaces forms the <u>cotangent bundle</u> T*M.

A tensor product of p 1-forms and q vectors

$$T^{(p,q)} = \omega^1 \otimes \ldots \otimes \omega^p \otimes X_1 \otimes \ldots \otimes X_q \tag{1.5}$$

is a <u>tensor</u> p times covariant and q times contravariant, or of type (q, p), and a general tensor of that type is a linear combination of the base tensors (1.5). The coefficients (with lower indices summed with 1-forms and upper indices summed with basis vectors) are the familiar tensors used traditionally in physics. One can similarly form the tensor spaces at each point of M and <u>tensor bundles</u>, and define tensor fields as sections of these bundles.

In the following section we take up in more detail the study of exterior forms, since they will play a more important role in the sequel and familiarity with exterior differential calculus will be useful.

<u>Remark.</u> Our notation in (1.5) is not quite standard. The 1-forms ω^i are usually denoted by $d\varphi^i$, since, as will become clear in sections 1.2, 1.3,one-forms are ideally suited for integration.

1.2. Exterior Calculus and Differential Forms

In this section we briefly review exterior algebra and discuss differential forms on a manifold. Section 1.3 sketches integration of differential forms on chains, recasting the familiar theorems of Newton-Leibniz, Green, Gauss, Stokes and Poincaré in a single formula for manifolds. The section ends with a discussion of distributions in terms of differential forms and a sketch of de Rham cohomology theory.

1.2.1. Exterior algebra. Let E be a real vector space (in particularly $\mathbb{R}^n$). The vectors in E will be denoted by lower-case latin letters: u, v, w,... . A 1-form on E is a linear form on E, i. e., a mapping $\omega: E \to \mathbb{R}$ such that

$$\omega(\alpha u + \beta v) = \alpha\omega(u) + \beta\omega(v), \quad \alpha,\beta \in \mathbb{R},\ u,v \in E.$$

We will also use the notation (1.4) for the value of a 1-form on a vector. It is obvious that 1-forms form a vector space E* and if E is n-dimensional, so is E*. Addition and multiplication by reals are defined as follows:

$$\langle \omega_1 + \omega_2, u\rangle = \langle \omega_1, u\rangle + \langle \omega_2, u\rangle \quad , \quad \langle\alpha\omega, u\rangle = \alpha\langle\omega, u\rangle .$$

A basis in the dual space will usually be denoted with upper indices: $\{\omega^i\}$, i = 1,..., n and the expansion of a 1-form in terms of this basis will be written with the summation convention as

$$\varphi = a_i\omega^i. \tag{1.6}$$

We see that 1-forms are essentially "covariant vectors" in terms of their components a_i.

The expansion (1.6) suggests a geometric interpretation of 1-forms: letting φ act on a vector u we get the right hand side of the equation of a hyperplane in E. Thus a 1-form can be pictured as a family of hyperplanes parallel to each other and its value on a vector counts how many such hyperplanes are intersected by this vector (for beautiful illustrations of this interpretation, cf. Misner-Thorne-Wheeler).

One physical example of a 1-form is the displacement dx in 3-space. The value of this 1-form on a force vector is the work done by the force in this displacement (usually identified with the scalar product F·dx).

Just as the displacement dx is a 1-form, an oriented area element dxdy is an example of a two-form. A 2-form (or exterior form of rank two) is a bilinear antisymmetric function of a pair of vectors:

$\omega^2: E \times E \to \mathbb{R}$:

$$\omega^2(\alpha u + \beta v, w) = \alpha\omega^2(u, w) + \beta\omega^2(v, w), \; \omega^2(u, v) = -\omega^2(v, u).$$

(In terms of indices, choosing a basis of 2-forms consisting of $n(n-1)/2$ forms, any two form is characterized by an "antisymmetric covariant tensor" of rank 2.)

The oriented area of a parallelogram spanned by two vectors u, v is bilinear and antisymmetric in u, v, hence a 2-form. Its value is given by the determinant of the components of the vectors u, v in an oriented orthonormal basis of the two-dimensional subspace spanned by the two vectors. Another example of a two-form is the flux of a fluid with velocity v through a parallelogram spanned by the vectors x, y. In vector notation this two-form is given by

$$\upsilon(x,y) = \underset{\sim}{v}\cdot(\underset{\sim}{x}\times\underset{\sim}{y}).$$

We shall see later that the "electromagnetic field strength tensor" is in fact a 2-form in Minkowski space.

We are now ready to define a p-form as a p-linear antisymmetric function from p vectors in E to real numbers $\omega: E \times E \times \ldots \times E \to \mathbb{R}$ such that:

$$\omega(\alpha u_1' + \beta u_1'', u_2, \ldots, u_p) = \alpha\omega(u_1', u_2, \ldots, u_p) + \beta\omega(u_1'', u_2, \ldots, u_p),$$
$$\omega(u_{i_1}, \ldots, u_{i_p}) = \text{sign } P\ \omega(u_1, \ldots, u_p), \tag{1.7}$$

where sign P, the signature of the permutation P which takes the indices $1, \ldots, p$ into $i_1, \ldots, i_p$, is +1 if P is even and -1 if P is odd.

The best known example of a three-form is the oriented volume of a parallelepiped spanned by three vectors u, v, w and equal to the determinant of the components of the three vectors in an orthonormal basis.

The set of all p-forms in an n-dimensional space, $p \leq n$, is a vector space of dimension $\binom{n}{p}$ (the binomial coefficient). In view of the antisymmetry, there cannot exist p-forms with $p > n$ in an n-dimensional vector space.

We now want to define an operation which will allow us to obtain two-forms out of one-forms, and in general, (p + q)-forms out of a p-form and a q-form. This operation is called <u>exterior multiplication</u> or <u>wedge-product</u> and is characterized by the following properties: we want

$$\omega^p \wedge \omega^q = \omega^{p+q} \tag{1.8}$$

to be

a) skew-symmetric (antisymmetric, in a way to be defined),

b) distributive with respect to addition,

c) associative.

One can proceed by induction, first defining the wedge product of two 1-forms on a pair of vectors as the determinant of the values of the two forms on the two vectors:

$$(\omega \wedge \varphi)(u, v) = \begin{vmatrix} \omega(u) & \varphi(u) \\ \omega(v) & \varphi(v) \end{vmatrix}. \tag{1.9}$$

<u>1.2.2. Definition.</u> The <u>wedge product</u> or <u>exterior product</u> of the p-form ω and the q-form φ is the p + q -form:

$$(\omega \wedge \varphi)(u_1,\dots,u_{p+q}) = \sum (-1)^{\nu}\omega(u_{i_1},\dots,u_{i_p})\,\varphi(u_{j_1},\dots,u_{j_q}),$$

where $i_1 < i_2 < \dots < i_p, j_1 < j_2 < \dots < j_q$ is a permutation of $1, 2,\dots, p + q$, the sum is over all permutations, and

$$\nu = \begin{cases} 0 \text{ if the permutation is even,} \\ 1 \text{ if the permutation is odd.} \end{cases}$$

It is easy to verify that the product of forms so defined is skew-symmetric, distributive and associative, where by skew symmetry we mean the following commutation property:

$$\omega \wedge \varphi = (-1)^{pq} \varphi \wedge \omega. \tag{1.10}$$

This shows that the wedge product of even forms is commutative, so that many operations on even forms are the same as on numbers (in particular, one can define determinants with entries 2-forms, fact which we will use later in defining characteristic classes).

1.2.3. Example. Let us consider a dynamical system with generalized coordinates $q_1,\ldots,q_n$ and generalized canonical momenta $p_1,\ldots,p_n$. We can then form the 2-form on $\mathbb{R}^{2n}$

$$\omega^2 = \sum p_i \wedge q_i, \tag{1.11}$$

which is the fundamental symplectic form of the dynamical system, playing an essential role in the construction of the Hamiltonian formalism (for details, cf. Abraham-Marsden and Arnol'd). The n-th exterior power of this form is proportional to the volume of the 2n-dimensional parallelepiped in phase space and by Liouville's theorem is a quantity conserved by the motion of the Hamiltonian system. Similarly, the lower exterior powers of this form are the various integral invariants of Poincaré and Cartan.

1.2.4. Using the wedge product we can form a basis in the space of p-forms by forming the exterior products of all possible p 1-forms (there are obviously $\binom{n}{p}$ linearly independent products of this type). Hence any p-form φ on the vector space $\mathbb{R}^n$ can be expanded in terms of the coordinate 1-forms ω^i, (1.6),

$$\varphi = \sum a_{i_1\ldots i_p} \omega^{i_1} \wedge \ldots \wedge \omega^{i_p}, \tag{1.12}$$

where the sum is over the increasing sequence of indices $i_1 < i_2 < \ldots < i_p$, ranging from 1 to n (we use the summation convention only when there is no restriction on the indices). The components $a_{i_1\ldots i_p}$ of the p-form are the value of φ on the basis vectors $e_{i_1},\ldots,e_{i_p}$.

1.2.5. If the basis vectors e_i are subjected to a linear transformation, the coefficients $a_{i_1 \ldots i_p}$ are subjected to the tensor product of the cogredient transformation, i. e., transform as a "covariant antisymmetric tensor" of rank p.

More generally, let $f: E \to F$ be a linear map of the vector space E into the vector space f and let ω be a p-form on F, Then the map f induces a p-form on E which is usually denoted by $f^*\omega$ (the upper asterisk shows that f^* is contragredient to the map f, or to the map f_* defined in (1.1)). The p-form $f^*\omega$ is defined by its values on p vectors $u_1, \ldots, u_p \in E$ in terms of the values of ω on their images under f:

$$(f^*\omega)(u_1, \ldots, u_p) = \omega(fu_1, \ldots, fu_p). \tag{1.13}$$

The form $f^*\omega$ is sometimes called the pullback of ω under f. It is clear that f^* is a linear mapping of the dual space F^* into the dual space E^*, and that choosing bases in E, F and the dual bases in F^*, E^*, the matrix of f^* will be the transpose of the matrix of f.

1.2.6. Differential forms. A 1-form on the tangent space T_xX of the manifold X at the point x is a linear mapping of T_xX into the real line $\mathbb{R}$. Given any differentiable real-valued function on X, the differential df_x of the function at the point x is such a 1-form, since it maps tangent vectors linearly into real numbers, Letting f be a smooth function on X, the differentials df_x are a smooth family of linear mappings of the tangent spaces at each point into the reals, i. e., a smooth mapping of the tangent bundle into the reals

$$df: TX \to \mathbb{R},$$

which is linear on each tangent space T_xX. We thus arrive at the following definition: A differential 1-form (or simply one-form) on the manifold X is a smooth mapping

$$\omega: TX \to \mathbb{R}$$

linear on each "fiber" (i. e., on each T_xX).

In other words, by analogy with the concept of vector field, a differential 1-form is a section of the cotangent bundle, i. e., a smooth mapping from the manifold X into the cotangent bundle T*X.

In $\mathbb{R}^n$ with coordinates $x_1,\ldots,x_n$ the components of the tangent vector at x are the values of the differentials $dx_1,\ldots,dx_n$ of the coordinates on that vector, so the dx_i can be considered as a basis of 1-forms. Consequently any differential 1-form on $\mathbb{R}^n$ with given coordinates can be written in the form

$$\omega = \sum \psi_i(x)dx_i$$

where the $\psi_i(x)$ are smooth functions. This explains the term differential form. (We remind the reader of the differentials in thermodynamics which are not exact differentials, or of the work done in a field of a nonconservative force .)

We are now naturally led to the definition of a differential p-form on a manifold as a p-form on the tangent space T_xX which is smooth in x, or in other words a section of the bundle of p-forms on TX.

1.2.7. Operations on differential forms. Addition and multiplication by real numbers are defined for differential forms pointwise, since at each point x of the manifold they are p-forms on the tangent vector space T_xX. It is obvious that these pointwise operations depend smoothly on the point and therefore yield again p-forms. Similarly, the wedge product of a differential p-form with a differential q-form will yield pointwise a (p + q)-form which depends smoothly on the point, and hence gives rise to a differential (p + q)-form. Moreover, any differential p-form can be multiplied by a smooth function (which at each point corresponds to multiplication by numbers), yielding a form of the same type. Thus the smooth differential forms are a module over the ring of real C^∞-functions on X. It is convenient to designate such smooth functions as 0-forms and to consider the multiplication of forms by 0-forms as a special case of exterior multiplication.

Since the differential of a smooth function is a differential 1-form we are led to the problem of generalizing the differentiation operator $d:f \mapsto df$ in such a manner that it produces a p + 1-form from a given p-form on a manifold. For this purpose we recall that on $\mathbb{R}^n$ a p-form could be expressed in terms of basis 1-forms by Eq. (1.12). We can achieve a similar decomposition of a differential p-form on a manifold X locally in terms of a chart. Let X be an n-dimensional manifold and $c = (U, \varphi, \mathbb{R}^n)$ a chart near the point x. Then a local basis of 1-forms is given by the differentials of the coordinate functions, $d\varphi^i$, $i = 1,\ldots,n$ (also called a frame of T^*X). An arbitrary differential p-form has the expansion

$$\omega = \sum_{i_1<\ldots<i_p} a_{i_1\ldots i_p}(x)\, d\varphi^{i_1} \wedge \ldots \wedge d\varphi^{i_p}, \tag{1.14}$$

where the $a_{i_1\ldots i_p}(x)$ are smooth functions of x. We then define the <u>exterior differential</u> of the form ω as the $(p + 1)$-form

$$d\omega = \sum da_{i_1\ldots i_p} \wedge d\varphi^{i_1} \wedge \ldots \wedge d\varphi^{i_p} \tag{1.15}$$

(summation as in (1.14)). The meaning of (1.15) can best be understood in terms of the exterior differential of the 1-form on $\mathbb{R}^3$

$$\omega = P dx + Q dy + R dz$$

with P, Q, R smooth functions of x, y, z. A simple calculation shows that

$$\begin{aligned} d\omega &= dP \wedge dx + dQ \wedge dy + dR \wedge dz \\ &= (D_x P dx + D_y P dy + D_z P dz) \wedge dx + (D_x Q dx + \ldots) \wedge dy \\ &\quad + (D_x R dx + \ldots) \wedge dz \\ &= (D_y R - D_z Q) dy \wedge dz + (D_z P - D_x R) dz \wedge dx \\ &\quad + (D_x Q - D_y P)\, dx \wedge dy, \end{aligned}$$

where $D_x P$, etc. denote the partial derivatives with respect to x, y, z. Thus, our calculation shows that the exterior differential of the 1-form $\omega = \underset{\sim}{V} \cdot d\underset{\sim}{r}$ is given by the 2-form curl $\underset{\sim}{V} \cdot d\underset{\sim}{\Sigma}$ where $d\underset{\sim}{\Sigma}$ is the oriented area element whose projections are $dx \wedge dy$, $dy \wedge dz$, $dz \wedge dx$. (The reader will have noticed that in three-space we associate to each 2-form a dual "axial vector", since a 2-form in three-space is characterized by three components; a 2-form in 4-space has six components, and is hence characterized by a pair of three-vectors, as is well known for the case of the electromagnetic field 2-form, cf. next section.)

The definition (1.15), or induction, shows that the exterior differential has the following properties:

i) the iterated exterior differential of any form vanishes:

$$d^2 = dd = 0;$$

ii) d is a "derivation" in the sense that if α is a p-form and β is a q-form on X, then

$$d(\alpha \wedge \beta) = d\alpha \wedge \beta + (-1)^p \alpha \wedge d\beta. \tag{1.17}$$

The property (1.16) follows from (1.15) by writing da in terms of partial derivatives:

$$\begin{aligned} dd\omega &= d[\sum_j \sum D_j a_{i_1 \ldots i_p} dx^j \wedge d\varphi^{i_1} \wedge \ldots \wedge d\varphi^{i_p}] \\ &= \sum\sum_{kj}\sum D_{kj} a_{i_1 \ldots i_p} dx^k \wedge dx^j \wedge d\varphi^{i_1} \wedge \ldots \wedge d\varphi^{i_p} = 0, \end{aligned}$$

on account of the symmetry of the second partial derivative $D_{kj} = D_{jk}$ and the antisymmetry of the exterior product. Property (1.17) is a consequence of the product rule of partial differentiation coupled with the skew-symmetry of the wedge product.

Now let $f: X \to Y$ be a morphism of the manifold X into the manifold Y. Then $f_*: TX \to TY$ is the tangent map, which is a linear map of $T_x X$ into $T_{f(x)} Y$. We can define the <u>pullback</u> of f as the map f^* which takes the p-form ω on Y into the p-form $f^*\omega$ on X by means of eq. (1.13) where $u_1, \ldots, u_p$ are p arbitrary vectors in $T_x X$:

$$(f^*\omega)(u_1, \ldots, u_p) = \omega(f_* u_1, \ldots, f_* u_p). \tag{1.18}$$

It is easy to see that

$$d(f^*\omega) = f^*(d\omega). \tag{1.19}$$

1.2.8. Examples. Consider ordinary 3-space with a curvilinear orthogonal coordinate system, such that the square of the line element has the form (for typing convenience we lowered the indices on $du^i = du_i$)

$$ds^2 = h_1^2\, du_1^2 + h_2^2\, du_2^2 + h_3^2\, du_3^2,$$

where the Lamé coefficients are smooth functions of u_1, u_2, u_3. Denoting by e_1, e_2, e_3 the unit vectors tangent to the coordinate lines we have a basis in the tangent space of our manifold at each point. Show that the values of the coordinate 1-forms du^i on the vectors e_j are given by $du^i(e_j) = h_i^{-1}\delta_j^i$. Thus, given a vector field in $\mathbb{R}^3$ with "contravariant" components $V = v^i e_i$, we can associate to V the 1-form given by the local inner product

$$\omega_V(e_i) = V\cdot e_i = v_i.$$

Hence $v_i = h_i v^i$ and the 1-form associated to the "covariant" components of V is (no summation over i in the preceding expression!):

$$\omega_V = v^1 h_1 du_1 + v^2 h_2 du_2 + v^3 h_3 du_3. \tag{1.20}$$

Similarly, we can associate to V the 2-form (axial vector):

$$\Omega_V = v^1 h_2 h_3 du_2\wedge du_3 + v^2 h_3 h_1 du_3\wedge du_1 + v^3 h_1 h_2 du_1\wedge du_2. \tag{1.21}$$

In particular, since the gradient of a smooth function f is a 1-form

$$df = D_1 f\, du_1 + D_2 f\, du_2 + D_3 f\, du_3,$$

we obtain for the components of the "gradient vector field" in the tangent space

$$(Vf)_i = h_i^{-1}\, D_i f \quad \text{(no summation)},$$

where $D_i f$ again denotes partial differentiation with respect to u_i. We leave it to the reader to specialize these results to the usual expressions in spherical and cylindrical coordinates.

Taking the exterior differential of (1.20) one obtains the expression of the 2-form curl V in curvilinear orthogonal coordinates:

$$\Omega_{\text{curl } V} = (D_2 v^3 h_3 - D_3 v^2 h_2)\, du_2 \wedge du_3 + \text{cycl. perm.}$$

which on comparison with (1.21) yields the components of the vector field curl V:

$$\text{curl } V = h_2^{-1} h_3^{-1} (D_2 v^3 h_3 - D_3 v^2 h_2) e_1 + \text{cycl. perm.}$$

Finally, the <u>divergence</u> of a vector field V can be defined in terms of the exterior differential of the 2-form (1.21) associated to it. Taking the exterior differential of (1.21) we obtain the 3-form

$$d\Omega_V = [D_1(v^1 h_2 h_3) + D_2(v^2 h_1 h_3) + D_3(v^3 h_1 h_2)]\, du_1 \wedge du_2 \wedge du_3 .$$

Since the local volume element is the exterior product of the three coordinate displacements $h_i du_i$ we define the divergence of V by

$$d\Omega_V = \text{div } V\, h_1 h_2 h_3\, du_1 \wedge du_2 \wedge du_3 ,$$

yielding the familiar expression for the divergence in curvilinear coordinates. Note that the divergence appears as a 3-form (one number).

<u>1.2.9. Definition</u>. A differential p-form ω is called <u>closed</u> if

$$d\omega = 0.$$

A p-form is called <u>exact</u> if there exists a (p - 1)-form φ such that it is the exterior differential of φ:

$$\omega = d\varphi .$$

Any exact form is closed ($d^2 = 0$), but the converse is not always true: the deviation of a closed form from exactness depends on the topological structure of the manifold. Loosely speaking, by counting the types of closed forms which are not exact one can find out how many "holes" of the appropriate dimension the manifold has. This idea will be discussed in more detail below, in the section on de Rham cohomology. Before this, we turn to a review of integration of differential forms on manifolds.

1.3. Integration of Differential Forms. Stokes' Theorem. De Rham Cohomology.

In this section we briefly outline without proofs the integration theory of forms and discuss Stokes' theorem, which unifies in a single formula the well known theorems of Newton-Leibniz, Gauss, Green and Stokes. We then give a sketch of the De Rham cohomology of forms. These results are not directly needed in the general exposition of gauge theory, and the reader may return to this section before studying characteristic classes. Some of the terms introduced here will occur in our discussion of electromagnetism in the next section.

1.3.1. Integration on chains. For a more detailed treatment of integration theory the reader may consult any of the listed textbooks on differential geometry; in particular Spivak has a simple and readable account.

Since on a manifold of dimension n any n-form is expressed by a number, one must suspect that all n-forms can be expressed in terms of a special one, which we call a volume element Ω, by multiplying the latter by a smooth function. Indeed, rewriting Eq. (1.14) for an n-form on $\mathbb{R}^n$ in the form

$$\omega = f(x)dx^1\wedge dx^2\wedge\ldots\wedge dx^n, \tag{1.22}$$

where $x^1,\ldots,x^n$ is a coordinate system (with an orientation), we obtain the natural volume element

$$\Omega = dx^1\wedge dx^2\wedge\ldots\wedge dx^n , \tag{1.23}$$

and the integral of the n-form ω over an oriented n-dimensional polyhedron P in $\mathbb{R}^n$ can be defined in terms of the ordinary Riemann integral (we use the standard notation $dx^1dx^2\ldots dx^n$ for the n-form Ω)

$$\int_P \omega = \int_P f(x)dx^1\ldots dx^n. \tag{1.24}$$

Here by orientation we mean the existence throughout $\mathbb{R}^n$ of a class of equivalent n-forms, two elements in the class differing from each other by a positive smooth function, and not vanishing anywhere. The reader will convince himself that this is equivalent to the usual definition of orientation, if he remembers that two volume elements in different coordinate systems differ from each other by a Jacobian and that they have the same orientation if the Jacobian is positive.

Another way of looking at integration is to remember that a differential form is a multilinear alternating form on the tangent space to the manifold at a given point. It will therefore yield a number when acting on an appropriate exterior product of linearly independent vectors. Thus, a 1-form ω^1 can be contracted with a vector to yield a number. If we now replace the vector by a path on the manifold (i. e. by a smooth mapping γ of the interval $[0,1]$ of the real line into M) we can subdivide the path γ by points ; take tangent vectors τ_k at these points and form the Riemann sum $\sum\omega(\tau_k)$; then go to the limit as the length of the intervals Δ_k in the parameter segment [0, 1] tend to zero. This yields the definition of the line integral of the 1-form:

$$\int\omega^1 = \lim_{\Delta_k\to 0} \sum_k \omega(\tau_k). \tag{1.25}$$

Similarly, if Ξ is a two-dimensional submanifold of our n-dimensional manifold M, one can define the integral of a 2-form ω^2 over Ξ by remembering that a two-form yields a number when applied to a bivector (the parallelogram spanned by two noncollinear vectors in the tangent plane to Ξ at a point; we remind the reader that a two-dimensional submanifold of M can be defined as the image under a smooth map from the **square** $[0,1]\times[0,1]$ in $\mathbb{R}^2$ to M). The image of a coordinate net in the square of the parameter space becomes a curvilinear coordinate net on the manifold Ξ and the bivectors tangent to the coordinate lines through a point give us the appropriate 2-cells on which the 2-form acts. One can again form a two-dimensional Riemann sum, go to the limit and define the integral.

In the general case of an m-dimensional cell in an n-dimensional manifold one proceeds in a similar manner. One takes an m-dimensional convex polyhedron (an m-vector) in $\mathbb{R}^m$ and maps it by a smooth function f into the manifold M, thus obtaining the cell in M. Then one defines the integral of the m-form ω over the cell $f(P)$ in M by the pullback $f^*\omega$ (1.18):

$$\int_{f(P)} \omega = \int_P f^*\omega .$$

It is easy to see that the integral so defined inherits all the properties of the right-hand side, i. e., is linear in the integrand and changes sign when the orientation of the cell changes. It is clear that the cell $f(P)$ defined by the smooth image of a polyhedron in M is a smooth submanifold of M only for very special functions f. Thus, even for a one-dimensional cell, the smooth image of a line segment, the set $f(D)$ may be a curve with self-intersections, or may degenerate into a point.

It is therefore convenient to consider in place of cells so-called <u>cell complexes</u>: a <u>chain</u> (of dimension m) on the manifold M consists of a finite number of oriented cells $f(P_i)$ which can be added and subtracted in the same manner as paths; thus, the cells can be multiplied by integers (positive, negative or zero) showing how many times and in which direction the cell is "traversed". Thus the chains of a given dimension become an abelian group, by simply adding the multiplicities. Among the chains on a manifold there is a distinguished subset, namely those chains which are the oriented <u>boundaries</u> of chains of the next higher dimension. If we denote a chain of dimension m by c_m, its boundary, a chain of dimension m - 1, is denoted by ∂c_m. One can obviously define an integral on the boundary of forms of degree m - 1. The integral of a form on a chain is defined as the linear combination with integer coefficients of the integrals over the cells which make it up, and similarly for the integral over the boundary.

<u>1.3.2. Examples.</u> Consider a vector field $V(x)$ in $\mathbb{R}^3$. To this vector field one can associate a one-form, namely the work done by the vector V in a displacement dx. The integral along an oriented curve C (which is a 1-chain in three-space) is the <u>circulation</u> of the vector field $V(x)$:

$$\int_C \omega = \oint_C V(x)\cdot dx. \tag{1.26}$$

The curve C has no boundary $\partial C = 0$, since it is itself the boundary of a two-dimensional cell (surface) S in $\mathbb{R}^3$. (It is a general property of the boundary operator that $\partial^2 = \partial\partial = 0$: a boundary of a boundary is zero.) Of course, the curve C may be the boundary of many smooth surfaces, but we consider a particular one, S.

The exterior differential of the 1-form ω is a 2-form (cf. the calculation following equation (1.15))

$$d\omega = \text{curl } V(x)\cdot dS,$$

where the vector dS is the oriented area element representing the 2-form with coordinates $dy\wedge dz$, $dz\wedge dx$, $dx\wedge dy$. We can integrate the form $d\omega$ over the 2-chain (here a cell) S:

$$\int_S d\omega = \iint_S V(x)\cdot dS. \tag{1.27}$$

By Stokes' theorem, well known from vector analysis, the right-hand sides of equations (1.26) and (1.27) are equal to one another. Consequently we also have equality of the left-hand sides, i. e. we can write Stokes' theorem in terms of differential 1-forms and their exterior differentials in the form

$$\int_S d\omega = \int_{\partial S} \omega. \tag{1.28}$$

We have also seen in Section 1.2.8 that to a vector field V(x) one can associate a 2-form (in $\mathbb{R}^3$ only), the flux of the vector field through an oriented area element dS:

$$\Omega = V(x)\cdot dS.$$

Let S be a closed oriented (in terms of the outward normal vector) surface, which is the boundary of a 3-chain (volume) D in $\mathbb{R}^3$. Then the integral of Ω over $S = \partial D$ is the <u>flux</u> through S:

$$\int_S \Omega = \iint_S V(x)\cdot dS \tag{1.29}$$

The exterior differential of Ω defines the divergence of V(x):

$$d\Omega = \text{div } V(x)\, dxdydz,$$

which can be integrated over the volume D:

$$\int_D d\Omega = \iiint_D \operatorname{div} V(x)\,dxdydz \qquad (1.30)$$

Now the right-hand sides of equations (1.30) and (1.29) are equal by Gauss' theorem, which in terms of the 2-form Ω takes a form which surprisingly is identical (mutatis mutandis) to (1.28):

$$\int_D d\Omega = \int_{\partial D} \Omega. \qquad (1.31)$$

1.3.3. Stokes' theorem for manifolds. A comparison of equations (1.28) and (1.31) with the fundamental theorem of calculus

$$\int_a^b df(x) = f(b) - f(a), \qquad (1.32)$$

where $f(x)$, a smooth function, is a 0-form and the points b, a form the boundary of the segment [a, b], with the appropriate orientation, leads us to conjecture the following generalized version of

Stokes' Theorem. (This theorem should rightfully be called, following Arnol'd, the Newton-Leibniz - Gauss-Green-Ostrogradskii-Stokes-Poincaré or NLGGOSP formula.) Let C be an arbitrary $m + 1$-chain on an n-dimensional manifold M, with m-dimensional boundary ∂M and let ω be an m-form on M with exterior differential $d\omega$. Then

$$\int_{\partial C} \omega = \int_C d\omega. \qquad (1.33)$$

We shall not give a proof here, referring the reader, e. g., to Spivak, but only mention that it suffices to prove it for a cell, which is the image of an oriented polyhedron in $\mathbb{R}^{m+1}$, with boundary consisting of m-planes parallel to the coordinate planes of a cartesian coordinate system. Then the integrations over the various coordinates reduce the proof to the fundamental theorem of calculus. Once the theorem is proved for a simple cell, it can be proved for an arbitrary chain, by decomposing the latter into cells and noting that the integrals over the interfaces cancel since each is traversed twice with opposite orientations, just as in the proof of, e. g., Green's theorem.

Remark. The preceding discussion of integration of differential forms on manifolds made the tacit assumption that either the manifold itself is compact, or that the chain on which the integration was carried out was compact (i. e., had finite "volume"), or finally, that the differential form itself had compact support (the support of a differential form is the set on which the form is different from zero). For differential forms of noncompact support, integrated over noncompact chains or manifolds, the integrals are to be understood as improper integrals, i. e., the "coefficient-functions" of the differential form (e. g., f(x) in (1,22)) must decrease sufficiently rapidly "at infinity". This leads us naturally to brief discussion of the notion of current.

1.3.4. De Rham Currents and Distributions. In the same manner that singular objects like the delta function and its derivatives (in $\mathbb{R}^n$) can be treated as distributions, i. e., as continuous linear functionals on appropriate test-function spaces (such as the space $\mathcal{D}$ of infinitely differentiable functions of compact support, or the space $\mathcal{S}$ of infinitely differentiable functions which together with all their derivatives fall off at infinity faster than any power of the distance from the origin), it makes sense to consider continuous linear functionals on vector spaces of smooth differential forms on a manifold, with support properties. Such linear functionals have been studied by de Rham who called them currents. Distributions on manifolds are special cases of currents, when the space of forms on which they are defined are 0-forms, i. e., smooth functions. Integrals and measures on manifolds are also special cases of currents. We only mention the basic definitions, and refer the reader for details to de Rham, or Dieudonné, vol III, Ch. 17.

Definition. Let X be an oriented manifold of dimension n. We denote by $\mathcal{D}_p(X;K)$ the complex vector space of smooth p-forms on X which have support in the compact subset K of X, and by $\mathcal{D}_p(X)$ the union of the spaces $\mathcal{D}_p(X;K)$ as K runs over X. When p = 0 we obtain the usual test-function space $\mathcal{D}(X)$ of smooth functions with compact support on X. A <u>p-current</u> (or <u>a complex p-current</u>) on X is by definition a linear functional T on $\mathcal{D}_p(X)$, which is continuous when restricted to each of the Fréchet spaces $\mathcal{D}_p(X;K)$ (where the seminorms on the latter are defined as usual, cf. Dieudonné). In other words, for each sequence of p-forms ω_i with supports in the same compact set K and which converges 0 (in the topology of $\mathcal{D}_p(X;K)$; this topology can be defined in terms of a chart by requiring uniform convergence to zero of all derivatives of the coefficient-function of the p-forms; this can be expressed more formally by seminorms, i. e., bounds on the absolute value of the coefficient-function and all its derivatives in K) the sequence $T(\omega_k) = \langle T, \omega_k \rangle$ converges to 0 in the complex plane.

The simplest example of a current is the Dirac p-current: let w be a tangent p-vector to X at the point x; the mapping which associates to a p-form ω its value on the p-vector w is linear and defines the Dirac current ε_{w_x}:

$$\varepsilon_{w_x}(\omega) = \langle \omega(x), w_x \rangle , \tag{1.34}$$

where the last expression is the value of the p-form on the p-vector at x. For p = 0 this reduces to the definition of the delta-function on a manifold (cf., e. g., Gel'fand-Shilov).

Another example is given by currents corresponding to locally integrable (n - p)-forms β on an n-dimensional <u>orientable</u> manifold X, with $0 \leq p \leq n$. For each p-form $\alpha \in \mathcal{D}_p(X;K)$ the n-form $\beta\wedge\alpha$ is locally integrable and has support in K, i. e., is integrable in K. The linear form $\alpha \to \int\beta\wedge\alpha$ on $\mathcal{D}_p(X)$ is continuous (this can be seen simply by introducing a chart and representing the forms α and β in terms of the canonical basis; cf. Dieudonné, Sec. (17.5.1)). This is reminiscent of the "generalized function" definition of distribution and we will, when necessary, use the integral notation even when the corresponding form is not locally integrable, e. g., for the Dirac current. We shall denote by T_β the current associated in this manner to the form β (note that T_β is a p-current, whereas β is an (n - p)-form, and that the orientation of X is essential here).

Further, if γ is a q-form where $q \leq p$, the (n - p + q)-form $\beta \wedge \gamma$ is locally integrable (even if γ is only continuous) and with the preceding notation we have

$$T_{\beta\wedge\gamma} = T_\beta \wedge \gamma. \tag{1.35}$$

The left-hand side is meaningful if γ is measurable and locally bounded, hence the right-hand side can be defined for such "bad" forms.

Finally, most of the known properties of distributions on $\mathbb{R}^n$ carry over to distributions (i. e., 0-currents) on manifolds. We refer the reader to the literature (Dieudonné, Gel'fand-Shilov) for details.

The space of all p-currents on a manifold X will be denoted by $\mathcal{D}'_p(X)$ and equipped with the weak topology. It will sometimes be convenient to think of a current as approximated by a weakly convergent sequence of currents of the type T_β or of currents of the type (1.35) which will be called regularizing sequences.

1.3.5. Closed and exact forms. De Rham cohomology. According to 1.2.9 a differential form is closed if its exterior differential is zero and is exact if it is the exterior differential of another form. The problem of characterizing closed forms, particularly, the problem of determining the properties of manifolds where closed forms are not exact is familiar to physicists in some special forms. Thus, the question: characterize a vector field such that curl V = 0, will elicit the answer V = grad f. But this answer is correct only if the domain of $\mathbb{R}^3$ in which V is defined is "star-shaped" with respect to the origin, or if this domain is simply connected in a two-dimensional sense, i. e., each loop can be contracted to a point (think of the magnetic field created by an infinite line current; it cannot be described by a single valued scalar potential, in spite of the fact that the curl vanishes everywhere outside the current). Similarly, the two-form in $\mathbb{R}^3 - 0$

$\omega = (xdydz - ydxdz + zdxdy)(x^2 + y^2 + z^2)^{3/2}$ (the origin was removed because this form has a singularity at the origin) is closed, but it is not the differential of a 1-form, i. e., it is not the curl of a vector field; this form measures the solid angle (or electric flux produced by a unit of charge at the origin) and its integral over the unit sphere yields 4π. We see from these two examples that the deviation of the forms from exactness "measures" the number and shape of the "holes" in the manifold on which they are defined. Thus, in the case of the curl which is not a gradient we find that space is not 2-connected, i. e. a line (carrying the sources) has to be eliminated, and in the second example the origin had to be eliminated.

Such properties were already known to Poincaré, who discovered a relation between the homology of manifolds and the number of closed forms which are not exact, but the general theory was developed in the 1930's by G. de Rham and has become known as *de Rham cohomology*. Here is a brief sketch of the basic concepts involved (for details, cf. de Rham, Greub-Halperin-Vanstone vol.I).

Let us denote the vector space of real (or complex) closed p-forms on a manifold M by $C^p(M; \mathbb{R})$ (or $C^p(M; \mathbb{C})$); such forms are also called p-cocycles on M. The subspace of closed forms which are exact (p-coboundaries) is denoted by $B^p(M; \mathbb{R})$ (respectively $B^p(M; \mathbb{C})$ in the complex case which we will need later). Since any vector space is an abelian group we can form the quotient group (we omit the $\mathbb{R}$ or $\mathbb{C}$ since the statements are valid in both cases)

$$H^p(M) = C^p(M)/B^p(M). \tag{1.36}$$

called the *p-th cohomology group* of the manifold. Its elements are equivalence classes of closed p-forms differing from each other by an exact form (or, using the other terminology: H^p consists of equivalence classes of cocycles, two cocycles being equivalent if they differ by a coboundary). The elements of $H^p(M)$ are the *p-th cohomology classes*. The dimension of $H^p(M)$ is a topological invariant of M called the p-th

Betti number b_p = dim $H^p(M)$ (this number is always finite if M is a compact manifold). The alternating sum

$$\chi_M = \sum_{p=0}^{n} (-1)^p b_p \qquad (1.37)$$

is called the Euler-Poincaré characteristic of the manifold. The Betti numbers and Euler-Poincaré characteristic can also be defined in a dual manner, by introducing triangulations of the manifold and the corresponding groups of chains and boundaries (the corresponding groups are called homology groups). The fact that the two definitions lead to the same Betti numbers is the content of de Rham's theorem.

Poincaré's lemma, stating that in a starshaped domain of $\mathbb{R}^n$ every closed form is exact can be restated as saying that if M is a starshaped domain of $\mathbb{R}^n$ then $H^0(M) \cong \mathbb{R}$ and $H^p(M) = 0$ for $p > 0$.

If we denote by S^n the n-dimensional sphere in $\mathbb{R}^{n+1}$ it can be shown that $H^0(S^n) \cong H^n(S^n) \cong \mathbb{R}$ and $H^p(S^n) = 0$ for $1 \leq p \leq n-1$. This implies that the Euler-Poincaré characteristic χ_{S^n} vanishes for n odd and equals 2 for n even. We will see later that it is related to the integral of the curvature of the sphere.

In the case of noncompact manifolds these definitions need some refining. One usually covers the manifold by charts based on sets with compact closure and discusses cochains and coboundaries associated to such coverings. For details (which will not be needed in the sequel) we refer the reader to the literature.

1.4. Vector-Valued and Lie-Algebra-Valued Differential Forms.

In the theory of connections, and in the discussion of nonabelian gauge theories we will require an extension of the concept of differential forms, which were defined as linear forms on the tangent space (more precisely, as sections of the cotangent bundle) with real values, to forms with values in a vector space, and in particular, in a Lie algebra. These will be defined as p-linear antisymmetric mappings from the tangent space to a vector space V, or a Lie algebra, respectively.

Let Λ_x^q denote the space of scalar-valued q-forms at the point x of the n-dimensional manifold M, i. e., the space of q-linear antisymmetric mappings of $\overset{q}{\wedge} T_xM$ (the space of q-vectors at x) into $\mathbb{R}$. Let V be an r-dimensional vector space. We shall call a V-valued differential q-form at x an element of the vector space $V \otimes \Lambda_x^q$, i. e., a q-linear antisymmetric mapping from T_xM into V.

Let a_j, j = 1, 2,...,r, be a basis of V. Then a vector-valued q-form Ξ can be uniquely represented in the form

$$\Xi = \sum_j a_j \otimes \xi^j = \sum a_j \xi^j, \quad \xi^j \in \Lambda_x^q. \tag{1.38}$$

This means that if $t_1,\ldots,t_q$ are tangent vectors in T_xM the value of the q-form $\Xi(x)$ on the q-vector $t_1 \wedge t_2 \wedge \ldots \wedge t_q$ will be

$$\Xi(x)\cdot(t_1 \wedge \ldots \wedge t_q) = \sum_j \langle \xi^j(x), t_1 \wedge \ldots \wedge t_q \rangle a_j. \tag{1.39}$$

We will use the same notation as in the left-hand side of (1.39) for the value of the q-form Ξ on vector fields $t_1, \ldots, t_q$. For any function F on M with values in V which is smooth

$$dF = \sum (dF^j) a_j \tag{1.40}$$

defines the differential, where $F = \Sigma F^j a_j$.

In terms of a coframe (i. e., a basis of Λ_x^q) $\eta^{\lambda_1} \wedge \ldots \wedge \eta^{\lambda_q}$ at x we can expand (1.38) further:

$$\Xi = \frac{1}{q!} \sum_j \sum_{\lambda_i} \cdots \sum \xi^j_{\lambda_1 \ldots \lambda_q} a_j \otimes \eta^{\lambda_1} \wedge \ldots \wedge \eta^{\lambda_q}, \tag{1.41}$$

where the coefficients are "a V-valued antisymmetric tensor".

The exterior differential of a vector-valued q-form can be defined by

$$d\Xi = \sum a_j d\xi^j, \tag{1.42}$$

can be easily seen not to depend on the basis and has the obvious properties of exterior differentiation.

However, since in general vectors cannot be multiplied, we have no exterior product of vector-valued differential forms.

This is no longer so if the vector space V is endowed with the structure of a Lie algebra, i. e., for any two vectors u, v $\in$ V there is a bracket [u, v], linear in each entry and with the usual properties:

$$[u, v] = -[v, u], \quad [u, [v, w]] + [v, [w, u]] + [w, [u, v]] = 0,$$
$$[a_i, a_j] = \sum c_{ij}{}^k a_k. \tag{1.43}$$

In this case we can define a bracket operation for the Lie-algebra valued p-form Ψ and q-form Ξ

$$\Psi = \sum a_j \psi^j, \quad \Xi = \sum a_k \xi^k, \quad j, k = 1,\ldots,r \tag{1.44}$$

by

$$[\Psi, \Xi] = \sum\sum [a_j, a_k] \otimes \psi^j \wedge \xi^k = \sum\sum\sum c_{jk}{}^i a_i \otimes \psi^j \wedge \xi^k. \tag{1.45}$$

From (1.45) it follows by a simple computation, taking into account the Jacobi identity (1.43), or the equivalent property of the structure constants $c_{ij}{}^k$, that

$$[\Psi, \Xi] = (-1)^{pq+1}[\Xi, \Psi], \tag{1.46}$$

and (for an s-form T)

$$(-1)^{ps}[\Psi,[\Xi,T]] + (-1)^{qp}[\Xi,[T,\Psi]] + (-1)^{sq}[T,[\Psi,\Xi]] = 0. \tag{1.47}$$

The formula for exterior differentiation of the bracket (the bracket plays the role of the wedge product for Lie-algebra valued forms) has the usual form:

$$d[\Psi, T] = (-1)^s[\Psi, dT] + [d\Psi, T]. \tag{1.48}$$

<u>1.4.1. Example. Maurer-Cartan forms on Lie Groups.</u> Let G be a Lie group, e its identity element. The Lie algebra g is the tangent space $T_e(G)$ at the identity element, with the bracket operation defined by the Lie bracket of the vector fields at e. As before, we denote a basis in $T_e(G) = g$ by (a_i), the base vector fields satisfying (1.43) (the standard notation is X_i, and sometimes the sign in (1.43) is reversed, i. e., $c_{ij}{}^k$ is replaced by $c_{ji}{}^k$; obviously $c_{ij}{}^k + c_{ji}{}^k = 0$). We denote a dual basis in $T_e^*(G) = g^*$ by (ω^j), $j = 1,\ldots,r = \dim g$:

$$\langle \omega^j, a_k \rangle = \delta_k^j. \tag{1.49}$$

If we denote left translations on the group by L_s, right actions by R_s

$$L_s t = st, \quad R_s t = ts, \quad s, t \in G; \quad L_a \circ R_b = R_b \circ L_a. \tag{1.50}$$

A tangent vector $X_e \in T_e(G)$ generates, by left translations (the derivative of the map L_s) a <u>left-invariant vector field</u> $X_s = L_{s*}X_e$.

Similarly, any 1-form $\omega_e \in T^*_e(G)$ gives rise to a left-invariant one-form ω_s, also called a Maurer-Cartan form on G, by the transposed map (pullback)

$$L^*_s\omega_s = \omega_e, \quad \omega_s = (L_{s^{-1}})^*\omega_e. \tag{1.51}$$

The linearly independent 1-forms ω^j of the basis (1.49) at e give rise to linearly independent 1-forms ω^i_s at the group element s by the action (1.51). Moreover, since the wedge products $\omega^j \wedge \omega^k$ form a basis for 2-forms at e, one can expand $d\omega^i$ in terms of these

$$d\omega^i = \sum\sum a^i_{jk}\omega^j\wedge\omega^k \quad , \quad a^i_{jk} + a^i_{kj} = 0. \tag{1.52}$$

In order to determine the coefficients a^i_{jk} we evaluate (1.52) on the basis elements a_m, $a_n \in T_e(G)$:

$$\begin{aligned} d\omega^i(a_m, a_n) &= \sum a^i_{jk}[\omega^j(a_m)\omega^k(a_n) - \omega^k(a_m)\omega^j(a_n)] \\ &= \sum a^i_{jk}(\delta^j_m\delta^k_n - \delta^k_m\delta^j_n) = a^i_{mn}. \end{aligned} \tag{1.53}$$

On the other hand (cf., e. g., Chevalley Ch.V, IV), for any vectors X, Y in g and any left-invariant 1-form ω, $d\omega(X, Y) = \tfrac{1}{2}\omega([X, Y])$, hence

$$\begin{aligned} d\omega^i(a_m, a_n) &= \tfrac{1}{2}\omega^i([a_m, a_n]) = \tfrac{1}{2}\sum c_{mn}{}^k\omega^i(a_k) = \tfrac{1}{2}\sum c_{mn}{}^k\delta^i_k \\ &= \tfrac{1}{2}c_{mn}{}^i, \end{aligned} \tag{1.54}$$

leading to the Maurer-Cartan structure equation

$$d\omega^i = \tfrac{1}{2}\, c_{jk}{}^i\omega^j\wedge\omega^k. \tag{1.55}$$

If we now consider Lie-algebra valued differential forms, as vector-valued differential forms with values in $T_e(G)$, we can write a left invariant g-valued 1-form which is independent of the choice of basis as

$$\omega = \sum (a_i)_s \otimes \omega^i_s \ , \tag{1.56}$$

which allows us to rewrite the Maurer-Cartan equation in coordinate-free form

$$d\omega = -\tfrac{1}{2}[\omega, \omega]. \tag{1.57}$$

In the special case when $G = GL(n, \mathbb{R})$, the group element is a real non-singular matrix. Then the Lie algebra is the space of all $n \times n$ matrices and the Maurer-Cartan form is $\omega = X^{-1}dX$, where $X \in G$.

2. ELECTROMAGNETISM AND DIFFERENTIAL GEOMETRY

2.0. Introduction. This chapter will illustrate some of the concepts introduced in the preceding chapter, particularly the concept of differential form and cohomology in the familiar context of electromagnetic theory.

Over the past century electromagnetic theory has repeatedly been subject to notational streamlining which has led to a deeper understanding of the concepts. One has only to compare Maxwell's original form of his equations in terms of components and coordinates (eight partial differential equations involving the components of E, B, D, and H, plus those involving the material) with the now standard vector form of these equations (this transition did not occur without resistance: those of us educated in the 40's still remember professors who presented vector analysis as something newfangled, to be used with a lot of caution). The four-dimensional tensor notation of special relativity has simplified the equations further, reducing them to two (if we restrict ourselves to vacuum equations). This has additionally led to a much deeper understanding of the nature of the electromagnetic field, reducing the distinction between electric and magnetic phenomena to changes in the frame of reference.

Tensor analysis became particularly important in curvilinear coordinates and in general relativity, but for general discussions and generalizations to other field theories it suffers from some of the disadvantages of Maxwell's original approach: sometimes one cannot see the physics hidden in the maze of tensor indices and coordinate transformation matrices. Fortunately Elie Cartan and his disciples have had the foresight to develop, starting in the 1920's and 1930's, the coordinate-free notation described in the preceding chapter. This notation has finally reached the practicing theoretical physicist starting in the late 1960's, and led to the streamlined form of Maxwell's equations presented in Chapters 3 and 4 of the Misner-Thorne-Wheeler treatise, to which we refer the reader for further details, illustrations and an exhaustive bibliography.

For the purposes of this chapter we will have to endow our manifold with a metric, in order to introduce the oncept of dual differential form and dual tensor. We will limit our attention to Minkowski space, only briefly mentioning how things change on a general manifold with Riemannian metric.

We will emphasize the concept of potential, which is apparently redundant, at least in a classical theory, but which will lead us to the main theme of these notes: gauge transformations, gauge theories and fiber bundles.

Finally, in order to illustrate the concept of de Rham cohomology introduced in the last chapter, we will briefly discuss the cohomological aspects of classical electromagnetic theory, which were recently emphasized by F. Strocchi [50] and discussed in a different context by J. Roberts [45].

2.1. Dual Forms and Maxwell's Equations. For the purposes of this chapter we will have to introduce the notion of dual of a differential form (or of a tensor). For this we will have to introduce the Levi-Civita symbol (totally antisymmetric tensor of maximal rank), which can only be done in an oriented manifold with metric. For most of the problems discussed here it will be sufficient to consider as our manifold Minkowski space, i. e., the manifold $\mathbb{R}^4$ endowed with the indefinite quadratic form $\eta(u, v)$ which to each two vectors associates a real number - the scalar product - with the usual properties except positive definiteness. In a basis where x^0 denotes the time and x^α ($\alpha = 1, 2, 3$) [1], the quadratic form η is characterized by the metric tensor η_{ij}:

[1] For typographical convenience we follow the convention of Landau and Lifshits, letting the latin indices from the middle of the alphabet take the values $i, j, k, \ldots = 0, 1, 2, 3$, and the greek indices from the beginning of the alphabet take the values: $\alpha, \beta, \gamma, \ldots = 1, 2, 3$.

$$\eta_{00} = 1, \eta_{11} = \eta_{22} = \eta_{33} = -1, \eta_{ij} = 0, i \neq j. \qquad (2.1)$$

Another concept related to the metric structure of space-time is the Levi-Civita tensor which can be considered as the tensor formed by the coefficients of the volume-element 4-form. In a particular Lorentz frame, with basis vector e_0 in the future time-direction (any direction inside the future light-cone will do) and e_1, e_2, e_3 forming a right-handed three-dimensional base (this fixes a standard orientation of the Minkowski manifold M) the Levi-Civita tensor is completely determined by its antisymmetry and the one covariant component:

$$e_{0123} = \langle \varepsilon, e_0 \wedge e_1 \wedge e_2 \wedge e_3 \rangle = +1, \qquad (2.2)$$

where ε denotes the 4-form of which the Levi-Civita tensor is the set of coefficients, and the middle expression denotes the evaluation of the 4-form on the "quadri-vector" spanning the volume element in M.

According to our discussion in the previous chapter, any 4-form is proportional to ε, in particular, the "volume element"

$$\begin{aligned} d^4\mathbf{x} &= e_{ijk\ell}dx^i dx^j dx^k dx^\ell \\ &= dx^0 dx^1 dx^2 dx^3. \end{aligned} \qquad (2.3)$$

It is easy to see that under a Lorentz transformation the tensor $e_{ijk\ell}$ gets multiplied by the determinant of the Lorentz transformation, and therefore, under a proper orthochronous Lorentz transformation (with determinant +1) it has the same components in any Lorentz frame with positive orientation. For other properties of the Levi-Civita tensor, cf. Misner-Thorne-Wheeler, Exercise 3.13 and Box 4.1,D, and Landau-Lifshitz, Sec.6,83). In particular, $e^{ijk\ell} = -e_{ijk\ell}$.

We remind the reader that the electromagnetic field can be described by the second-rank antisymmetric tensor (2-form)

$$\begin{aligned} F = \tfrac{1}{2} \sum_{ij} F_{ij}dx^i \wedge dx^j &= F_{10}dx\wedge dt + F_{20}dy\wedge dt + F_{30}dz\wedge dt \\ &\quad + F_{23}dy\wedge dz + F_{31}dz\wedge dx + F_{12}dx\wedge dy \\ &= E_x dx\wedge dt + E_y dy\wedge dt + E_z dz\wedge dt + B_x dy\wedge dz + B_y dz\wedge dx + B_z dx\wedge dy, \end{aligned} \qquad (2.4)$$

which establishes the identification of the pair of 3-vectors $(\underset{\sim}{E}, \underset{\sim}{B})$

with the components of the "Faraday" tensor F_{ij} (to use the terminology advocated by Misner-Thorne-Wheeler), or the coefficients of the two-form F in a coordinate basis. In terms of the two-form F it is easy to see that the homogeneous pair of Maxwell equations:

$$\operatorname{div} \underset{\sim}{B} = 0, \quad \dot{\underset{\sim}{B}} + \operatorname{curl} \underset{\sim}{E} = 0, \tag{2.5}$$

where the dot denotes partial derivative with respect to time and c has been set equal to one, is equivalent to the single exterior differential equation

$$dF = 0, \tag{2.6}$$

which says that F is an exact two-form. Eq. (2.6) can be rewritten in terms of components as

$$\frac{\partial F_{ij}}{\partial x^k} dx^i \wedge dx^j \wedge dx^k = 0 \text{ (summation convention!)}, \tag{2.7}$$

which in turn can be written as the set of tensor equations:

$$\frac{\partial F_{ij}}{\partial x^k} + \frac{\partial F_{jk}}{\partial x^i} + \frac{\partial F_{ki}}{\partial x^j} = 0, \tag{2.8}$$

obviously equivalent to the four equations (2.5).

The inhomogeneous pair of Maxwell equations

$$\operatorname{div} \underset{\sim}{E} = \rho, \quad \dot{\underset{\sim}{E}} - \operatorname{curl} \underset{\sim}{B} = -\underset{\sim}{j}, \tag{2.9}$$

(we have used Lorentz-Heaviside units to avoid the factors of 4π in the right-hand sides, c = 1), reduces to the tensor equation

$$\frac{\partial F^{ij}}{\partial x^j} = J^i \text{ (summation convention!)}, \tag{2.10}$$

where the four-vector J^i has components $J^0 = \rho$, $(J^1, J^2, J^3) = \underset{\sim}{j}$. In order to rewrite this last equation in coordinate-free form we are forced to use the Levi-Civita tensor to define the <u>dual</u> of the Faraday tensor (called by Misner-Thorne-Wheeler) the <u>Maxwell</u> tensor) and of J^i

$$*F_{ij} = \tfrac{1}{2} F^{k\ell} e_{k\ell ij}, \quad *J_{ijk} = J^{\ell} e_{\ell ijk}, \tag{2.11}$$

which define respectively the 2-form *F and the 3-form *J:

$$*F = -B_x dx\wedge dt - B_y dy\wedge dt - B_z dz\wedge dt + E_x dy\wedge dz + E_y dz\wedge dx + E_z dx\wedge dy,$$
$$*J = \rho\, dx\wedge dy\wedge dz + j_x dy\wedge dz\wedge dt + j_y dz\wedge dt\wedge dx + j_z dt\wedge dx\wedge dy. \tag{2.12}$$

The latter has a simple interpretation if we integrate the 3-form $*J$ over a three-dimensional spacelike surface S; in that case the last three terms do not contribute since $dt = 0$, and we obtain

$$\int_S *J = \int_S \rho\, dx\wedge dy\wedge dz = Q(S), \qquad (2.13)$$

i. e., the total charge contained in the 3-dimensional region S.

The inhomogeneous pair of Maxwell equations reduces to the single exterior differential equation for the 2-form $*F$:

$$d{*F} = *J. \qquad (2.14)$$

Eq. (2.14) has an immediate consequence: since the 3-form $*J$ is exact, i. e., the exterior differential of the 2-form $*F$, it is automatically closed, i. e.,

$$dd{*F} = d{*J} = 0. \qquad (2.15)$$

But this equation is nothing other than the continuity equation for the 4-current J^i (this is easily seen by writing (2.15) in terms of the expansion (2.12) and factoring out the volume element d^4x).

Combining Eq. (2.15) with Stokes' theorem applied to a space-time region V bounded by the two spacelike 3-surfaces S_1 and S_2 which form the boundary of V (and which may be assumed to coincide at sufficiently large spacelike distance, thus enclosing a finite 4-volume):

$$0 = \int_V d{*J} = \int_{\partial V} *J = Q(S_2) - Q(S_1), \qquad (2.16)$$

we obtain the global form of the law of conservation of electric charge.

Finally, since F is a closed 2-form, and Poincaré's lemma applies to a region of Minkowski space which does not contain point-charges (i. e., a region where F is a smooth 2-form), F itself must be the exterior differential (curl) of a 1-form A:

$$F = dA. \qquad (2.17)$$

The coefficients of the 1-form A form the covariant vector-potential A_i: $A = A_i dx^i$, which is itself determined only up to the addition of the differential (gradient) of a 0-form (smooth function) f: A and $A + df$ determine the same field F. Substituting (2.17) into (2.14) and taking the dual (remembering that the dual of a dual leads back

back to the original tensor with changed sign) we obtain the "wave equation for the vector potential" (the 1-form dual to the 3-form $*J$)

$$*d*F = *d*dA = J; \tag{2.18}$$

written out in components this equation contains beside the D'Alembertian of the potential A_i also the x^i-derivative of the divergence $\partial_j A^j$, which is usually set equal to zero (Lorentz condition). We leave it for the reader to translate the Lorentz condition into the language of differential forms (hint: differentiate the 3-form $*A$ dual to A and take the dual).

We are now ready to introduce the following general definition.

2.1.1. Definition. In an n-dimensional space with metric the dual of a p-form ω is the (n - p)-form $*\omega$ with the components:

$$(*\omega)_{j_1\ldots j_{n-p}} = (1/p!)\,\omega^{i_1\ldots i_p}\, e_{i_1\ldots i_p j_1\ldots j_{n-p}},$$

where $e_{i_1\ldots i_n}$ is the totally antisymmetric tensor of n-space, which can be considered the coefficient of the volume element n-form for the standard orientation of n-space. We denote this form by ε.

The dual of a dual of a p-form is equal to that form if the form is odd, and equal to its negative when the form is even, i. e.,

$$**\omega = (-1)^{p-1}\omega. \tag{2.19}$$

The wedge-product of a p-form with its dual is an n-form, hence proportional to the volume element n-form ε . The coefficient of ε is defined to be the norm of the p-form:

$$\omega \wedge *\omega = \|\omega\|^2\varepsilon \qquad \|\omega\| = (1/p!)\,\omega_{i_1\ldots i_p}\,\omega^{i_1\ldots i_p}. \tag{2.20}$$

(In all these equations we have used the summation convention and the factor 1/p! takes into account the combinatorics; alternatively, one can sum over ordered sets of indices, which are all different, and omit the factorial, as is done in Misner-Thorne-Wheeler, to which the reader is again referred for many of the details and illustrations.)

Example. From the 2-forms F and $*F$ one can form the invariants:

$$*(F \wedge *F) = \|F\|^2 = \underset{\sim}{B}^2 - \underset{\sim}{E}^2, \quad \underset{\sim}{E}\cdot\underset{\sim}{B} = \tfrac{1}{4}F_{ij}{*F}^{ij} = F \wedge F(\varepsilon),$$

where the expressions in terms of the field vectors should be understood as multiplied with a volume element, i. e., as densities, with a loose meaning of the equality signs. The first of these invariants is the Lagrangian density of the electromagnetic field.

2.1.2. Generalization to Riemannian manifolds. A Riemannian manifold is a manifold in which a metric (definite or not) is defined in each chart and is assumed to vary smoothly over the manigold. We denote such a bilinear symmetric form by g (components g_{ij} in a chart), and assume that the reader is familiar with the basic mechanics of raising and lowering indices. Here we restrict our attention to 4-dimensional manifolds. We remind the reader of the definitions of volume element and Levi-Civita tensor (or density) on a Riemannian manifold (cf., e. g., Landau-Lifshits, § 83, or Misner-Thorne-Wheeler, §§ 8.4, 22.4).

We denote the Galileean coordinates in a given chart, in which the metric tensor g_{ij} reduces to its Minkowski (or Euclidean) form η_{ij} by x'^i, and coordinates in an arbitrary chart by x^i. The transformation of a vector from the Galileean chart to the arbitrary chart is given by the matrix $(\partial x^i/\partial x'^j)$, hence the contravariant antisymmetric object e^{ikmn} transforms into (we follow the Landau-Lifshits notation):

$$E^{ikmn} = \frac{\partial x^i}{\partial x'^p}\frac{\partial x^k}{\partial x'^r}\frac{\partial x^m}{\partial x'^s}\frac{\partial x^n}{\partial x'^t}e^{prst}. \tag{2.21}$$

The antisymmetric sum of products of derivatives in this equation is nothing else than the Jacobian of the transformation from the galileean chart to the curvilinear coordinate chart:

$$J = \frac{\partial(x^0, x^1, x^2, x^3)}{\partial(x'^0, x'^1, x'^2, x'^3)} \tag{2.22}$$

and the latter is equal to

$$J = 1/\sqrt{-g} = 1/\sqrt{|g|}. \tag{2.23}$$

Here g denotes the determinant of the tensor g_{ij} (the determinant of η_{ij} is -1, hence the minus sign, or the absolute value, which is preferable, since it remains valid also in the case of positive definite metric). Hence:

$$E^{ikmn} = |g|^{-\frac{1}{2}}e^{ikmn}, \quad E_{ikmn} = |g|^{\frac{1}{2}}e_{ikmn}. \tag{2.24}$$

In general coordinates the volume element (and similarly other, lower-dimensional forms which act as integration elements) can be represented either as d^4x multiplied by the appropriate Jacobian, or with the help of the Levi-Civita tensor E_{ikmn}:

$$d^4x' \to \sqrt{|g|}\, d^4x = (1/4!)E_{ikmn}dx^i \wedge dx^k \wedge dx^m \wedge dx^n. \qquad (2.25)$$

The duals of tensors and differential forms are to be defined in terms of the E_{ijkm} (and are therefore affected by a $\sqrt{-g}$). In particular, the volume element (2.25) may be considered to be the dual of the 0-form 1, and the 4-form dual to the 0-form (smooth function) f is that function multiplied by the volume element (2.25). Thus,

$$\int *f = \int f\sqrt{-g}\, d^4x, \qquad (2.26)$$

is the volume integral of the function f.

The integral

$$(\alpha,\beta) = \int \alpha\wedge *\beta = (\beta, \alpha) \qquad (2.27)$$

of two p-forms defines an inner product among p-forms, if the integral converges (e. g., if the region of integration is compact, or one of the two forms has compact support). This allows one to introduce a dual of the operator d, acting on dual forms, and denoted by δ:

$$(d\alpha, \beta) = (\alpha, \delta\beta), \qquad (2.28)$$

which plays the role of the adjoint of d. It is easy to see that in an even-dimensional manifold, in particular, in spacetime,

$$\delta\alpha = -*d*\alpha, \qquad (2.29)$$

for any p-form, i. e., the adjoint of d with respect to the inner product (2.27) is $-*d*$. The "second-order" operator which generalizes the Laplacian (for Riemannian manifolds) or the D'Alembertian (for the pseudoriemannian case discussed here) is the <u>de Rham - Lichnerowicz operator</u> (for the Laplacian this is known as the <u>Hodge decomposition</u>):

$$\Box = d\delta + \delta d = -d*d* -*d*d, \qquad (2.30)$$

which in the case of 0-forms reduces to the ordinary D'Alembertian in curvilinear coordinates, and for the case of 4-vectors can be found, e. g., in de Rham §26, or Misner-Thorne-Wheeler, Eq. (22.19d).

In order to transcribe the equations of electrodynamics into curved spacetime (or curvilinear coordinates in flat spacetime) one usually replaces the ordinary derivatives in the equations by covariant derivatives. One then shows that both in the expression of F in terms of A and in the equations corresponding to our shorthand dF = 0, the antisymmetry allows one to replace covariant by ordinary derivatives (cf., e. g., Landau-Lifshits, §90, or Misner-Thorne-Wheeler, § 22.4). However, the inhomogeneous Maxwell equation (2.18), has the more complicated coordinate form:

$$\partial_j(\sqrt{-g}F^{ij}) = \sqrt{-g}\, J^i, \qquad (2.31)$$

where ∂_j denotes the partial derivative with respect to x^j. If we remember that the $(-g)^{\frac{1}{2}}$ is part of the definition of dual forms, it is easy to see that Eq. (2.31) is the coordinate form of the equation

$$d{*F} = {*J} \qquad (2.32)$$

(we leave the proof as a simple exercise to the reader). Further, replacing F by dA, and taking duals, we obtain the de Rham form for the wave equation if we impose the Lorentz gauge condition (this time with covariant 4-divergence!).

The contents of this subsection will not be needed in the sequel, but were meant only to show that the presentation of electromagnetism in terms of differential forms remains unchanged in general coordinates, whereas the component form becomes more complicated.

2.2. The Potential 1-Form and Gauge Transformations of Charged Particle Fields. In this section we return to a more detailed study of the one-form A by which we have replaced the 4-vector potential, noting that it appears naturally if one takes a variational principle as the starting point for the description of the motion of a charged particle in an electromagnetic field. As an historical curiosity it is worth mentioning that this variational approach was introduced by R. Schwarzschild two years before Einstein wrote his first paper on relativity (R. Schwarzschild, Zur Elektrodynamik, Göttinger Nachr., 1903, quoted in Sommerfeld. The only difference between Schwarzschild's formulation and the present one (which essentially translates the standard treatment, to be found, e. g., in Landau-Lifshits, § 16, into coordinate-free notation) is his use of the nonrelativistic Lagrangian for the particle motion.

We remind the reader that the motion of a free particle of mass m in Minkowski space is geodesic, i. e., occurs along extremals of the action integral

$$S = -m\int_a^b ds = -m\int_{t_a}^{t_b} (1 - v^2)^{\frac{1}{2}} dt, \qquad (2.33)$$

where the integral is taken along the world line of the particle between the events a and b (corresponding to the times t_a, t_b), v is the velocity of the particle and $ds = (\eta_{ij} dx^i dx^j)^{\frac{1}{2}}$, c = 1 . If the particle now moves in an electromagnetic field, one must add to the action (2.33) a term describing the interaction of the particle with the field. One can consider the following reasoning as a combination of the known empirical behavior of particles (e. g., the fact that the potential energy of a particle of charge e in an electrostatic field described by the potential φ is $e\varphi$, or that the addition to the nonrelativistic Lagrangian in a magnetic field is given by $e\underset{\sim}{A}\cdot\underset{\sim}{v}$ where $\underset{\sim}{A}$ is the vector potential) with the fact that the only "reasonable" invariant one can add to the line integral (2.33) which carries with it information about the electromagnetic field and is "simple", i. e., leads to the known equations of motion, is a line integral of the one-form A which corresponds to the 4-vector potential. Hence the action for a particle moving in a given electromagnetic field has the form

$$S = -\int_a^b (m ds + eA) = -\int_a^b (m ds + eA_j dx^j) . \qquad (2.34)$$

The three-momentum 1-form is defined as the gradient of the Lagrange function with respect to the velocity, yielding the well-known form

$$P = (p_\alpha + eA_\alpha) dx^\alpha = P_\alpha dx^\alpha \quad (= 1, 2, 3) . \qquad (2.35)$$

Since one can add to any line integral an exact differential (i. e., an exact one-form) without changing the equations of motion, it is obvious from (2.34) that the 1-form A is determined only up to an exact 1-form $d\chi$, where χ is a 0-form, i. e., a smooth function. The transformation

$$A \rightarrow A + d\chi \tag{2.36}$$

is a <u>gauge transformation</u> of the potential 1-form A, which does not affect the equations of motion of the charge. Indeed, it is easy to show that the Euler-Lagrange equations of the action (2.34) (after some vector manipulations leads to the equation of motion for the 4-momentum 1-form $p = p_i dx^i$:

$$\frac{dp}{ds} = e\langle F, u\rangle \text{ or } \frac{dp_i}{ds} = eF_{ij}u^j, \tag{2.37}$$

where $F = dA$ is the electromagnetic field 2-form (2.4) and $\langle\ ,\ \rangle$ denotes the evaluation of the 2-form F on the 4-vector u (the four-velocity $u^i = dx^i/ds$), yielding a 1-form. Since $dd\chi = 0$, the gauge transformation (2.36) leaves the right-hand side of the equation of motion (2.37) unchanged, and one can argue that at the level of classical particle mechanics the potential 1-form is a redundant, though convenient, quantity. We will see that this is no longer so in a quantum theory.

To complete the picture we have to add to the action (2.34) the action of the electromagnetic field, the variation of which with respect to A_i would lead to Maxwell's equations. Here we will be guided by two principles, namely invariance and simplicity, the first meaning that the field action integral must be the integral of a 4-form (otherwise the action will not be invariant under Lorentz transformations), and that the 4-form used should be such as to lead, in the absence of charged particles to the homogeneous linear partial differential equations obtained from Maxwell's equations when $j^i = 0$. This means that we have to look for a 4-form depending on the field variables, which is at most quadratic in field variables. In addition, it is important to remember that one can always add to the action integrand an exact 4-form, i. e., the exterior differential of a 3-form.

The only 4-forms which are quadratic in field quantities which are (we also give their coordinate expressions, as well as the expressions in terms of $\underset{\sim}{B}$ and $\underset{\sim}{E}$):

$$F\wedge *F = \tfrac{1}{2}F_{ij}dx^i\wedge dx^j\wedge\tfrac{1}{2}*F_{mn}dx^m\wedge dx^n = \tfrac{1}{4}F_{ij}*F_{mn}e^{ijmn}d^4x$$
$$= \tfrac{1}{2}F_{ij}F^{ij}d^4x = \quad (B^2 - E^2)d^4x. \tag{2.38}$$

and

$$F\wedge F = \tfrac{1}{2}F_{ij}dx^i\wedge dx^j\wedge\tfrac{1}{2}F_{mn}dx^m\wedge dx^n = \tfrac{1}{2}F_{ij}*F^{ij}d^4x = \underset{\sim}{E}\cdot\underset{\sim}{B}d^4x. \tag{2.39}$$

In addition to being a pseudoscalar (i. e., changing sign under improper Lorentz transformations) the second 4-form is in fact exact, as can easily be seen from the following string of identities:

$$F\wedge F = dA\wedge dA = d(A\wedge dA) - A\wedge ddA = d(A\wedge dA), \tag{2.40}$$

since $d^2A = ddA = 0$. Consequently, adding a term proportional to (2.40) to an integrand in the action is tantamount to adding a surface term (by Gauss', or Stokes' theorem), i. e., at most modifies the boundary conditions and has the same effect as a gauge transformation.

Therefore, the only Lagrangian density one can form out of the electromagnetic field quantities is proportional to (2.38), and an easy calculation (or the reduction to the three-dimensional form) shows that the proportionality factor must be $-\tfrac{1}{2}$; hence the action of the electromagnetic field is

$$S_F = -\tfrac{1}{2}\int F\wedge *F = -\tfrac{1}{4}\int F_{ij}F^{ij}d^4x = \tfrac{1}{2}\int(E^2 - B^2)d^4x. \tag{2.41}$$

This term has to be added to (2.34) in order to obtain the total action of particles plus field. Note that the integral in (2.41) is over spacetime, whereas that in (2.34) is a line integral over the world line of the particle.

One can obtain Maxwell's equations by varying the total action with respect to A_i, provided we replace the line current density edx^i in (2.34) by an arbitrary current-density four-vector j^i, and in turn replace the latter with the dual 3-form $*J$, so that the field-particle action becomes

$$S_{FP} = -\int A\wedge *J = -\int A_i j^i d^4x. \tag{2.42}$$

The line integral in (2.34) corresponds to a point charge described by a delta-function charge density $\rho = e\delta(\underset{\sim}{x} - \underset{\sim}{x}_i)$, $j^i = dx^i/dt$, where $\underset{\sim}{x}_i$ denotes the position of the particle with label i.

The action principle for a charged field, for the Schrödinger (or Dirac) wave function, or for a quantized particle field is then obtained by adding to the terms (2.41) and (2.42) the action integral for the field under consideration, which is again constructed (up to an exact 4-form, or divergence) by using arguments of invariance, simplicity, or, when possible, so that one obtains a desired wave equation for the appropriate field.

For simplicity we consider the case of a complex scalar field φ, with complex conjugate (or hermitian adjoint, in the quantized case) φ^*. The simplest action integral for such a field, satsifying the requirment that it lead to a linear partial differential equation of order at most two (in fact, the Klein-Gordon equation) has the form:

$$S_\varphi = \int (\partial_i \varphi^* \partial^i \varphi - m^2 \varphi^* \varphi) d^4x$$

$$\simeq -\int \varphi^* (\partial_i \partial^i \varphi + m^2 \varphi) d^4x. \qquad (2.43)$$

Here the second integral differs from the first by a divergence, fact which we have denoted by the sign $\simeq$. Varying the free-field action (2.43) with respect to φ^* yields the Klein-Gordon equation (after an integration by parts for the first integral, or directly for the second integral).

In addition to being Lorentz invariant, the integral (2.43) is also invariant under the group U(1), where χ is a parameter,

$$\varphi(x) \to e^{ie\chi}\varphi(x), \ \varphi^*(x) \to e^{-ie\chi}\varphi^*(x). \qquad (2.44)$$

According to Emmy Noether's theorem on variational problems with invariance properties this implies the existence of a "conserved current"

$$j_k = ie(\varphi^* \partial_k \varphi - \partial_k \varphi^* \varphi); \quad \partial_k j^k = 0, \qquad (2.45)$$

where e is a "charge". This current can be coupled to the one-form A in the manner shown in Eq. (2.42), thus leading to the interpretation of the fields φ, φ^* as describing a charged field of particles and antipaticles (this interpretation is valid in a quantum theory).

If we now add (2.41), (2.42) and (2.43) we obtain an action for the coupled scalar and electromagnetic fields, which in addition to the invariance properties which are obvious (Lorentz invariance, U(1)-invariance (2.44) is also invariant under a more general gauge transformation of the second kind, or local gauge transformation, where t the parameter χ in Eq. (2.44) is replaced by a smooth function $\chi(x)$, and simultaneously the potential one-form A is subjected to the gauge-transformation (2.36), with the same "gauge function" χ:

$$\varphi(x) \to e^{ie\chi(x)}\varphi(x), \quad \varphi^*(x) \to e^{-ie\chi(x)}\varphi^*(x),$$

$$A \to A + d\chi. \tag{2.46}$$

This statement is not quite correct, since the use of the action (2.43) for the scalar field leads, under the gauge transformation (2.46) to the appearance of an additional term in the current (2.45), namely a term $\varphi^*\varphi A_i$ which has to be multiplied by the charge squared, e^2, and corresponds to a nonlinear term in the resulting wave equation. In the following section we will remedy this, by replacing the action for the field φ by an action integral leading to a system of first order partial differential equations equivalent to the Klein-Gordon equation. Similar results hold for the Dirac equation and for the Schrödinger equation. In the latter, a closer analysis of the gauge transformation (2.46) led to the discovery of the Bohm-Aharonov effect, an interference effect showing that in a quantum mechanical context the potential one-form is not completely redundant (for a discussion of the interpretation of this effect, cf. Wu and Yang [62] and Strocchi and Wightman [51]).

2.3. Hermann Weyl's Gauge Principle and the Yang-Mills Generalization to Nonabelian Groups. We now elevate the idea of gauge transformations which depend on the point to a principle which allows us to "reinvent" electromagnetism. This point of view is originally due to Hermann Weyl [61], was extensively used by Schwinger [48] in his general theory of quantized fields, and was extended by Yang and Mills [63] to fields invariant under the group SU(2), leading to the development of the theory of nonabelian gauge fields. We will see that the analogy of local gauge transformations to changes of frames in the tangent bundle of a manifold will lead us naturally to the various fiber-bundle concepts which will be developed in the following chapter.

As a working example we again consider a complex scalar field φ, but in order to avoid the problem posed by defining the current in the presence of the electromagnetic field (mentioned after Eq. (2.46)) and in order to have a formalism which extends immediately to the Dirac field and other multi-component fields, we replace the second-order Klein-Gordon equation by an equivalent system of first-order equations (a special form of the Duffin-Kemmer-Petiau equations). Most of the discussion of this section is valid both in the classical and quantum theories, except that in the latter the ordering of field operators may become important. We do not pay attention here to this problem.

Thus, in place of the scalar field φ we introduce the five fields

$$u_0 = (i/m)\partial_0\varphi, \ \ldots \ , u_3 = (i/m)\partial_3\varphi, \quad u_4 = \varphi, \tag{2.47}$$

and the six 5 × 5 matrices (these are a special representation of the Duffin-Kemmer matrices)

$$\beta^0 = \begin{pmatrix} 0 & 0 & 0 & 0 & -1 \\ 0 & 0 & 0 & 0 & 0 \\ 0 & 0 & 0 & 0 & 0 \\ 0 & 0 & 0 & 0 & 0 \\ 1 & 0 & 0 & 0 & 0 \end{pmatrix}, \qquad \beta^1 = \begin{pmatrix} 0 & 0 & 0 & 0 & 0 \\ 0 & 0 & 0 & 0 & -1 \\ 0 & 0 & 0 & 0 & 0 \\ 0 & 0 & 0 & 0 & 0 \\ 0 & 1 & 0 & 0 & 0 \end{pmatrix},$$

$$\beta^2 = \begin{pmatrix} 0 & 0 & 0 & 0 & 0 \\ 0 & 0 & 0 & 0 & 0 \\ 0 & 0 & 0 & 0 & -1 \\ 0 & 0 & 0 & 0 & 0 \\ 0 & 0 & 1 & 0 & 0 \end{pmatrix}, \qquad \beta^3 = \begin{pmatrix} 0 & 0 & 0 & 0 & 0 \\ 0 & 0 & 0 & 0 & 0 \\ 0 & 0 & 0 & 0 & 0 \\ 0 & 0 & 0 & 0 & -1 \\ 0 & 0 & 0 & 1 & 0 \end{pmatrix}, \tag{2.48}$$

$$\beta^4 = \begin{pmatrix} 1 & 0 & 0 & 0 & 0 \\ 0 & -1 & 0 & 0 & 0 \\ 0 & 0 & -1 & 0 & 0 \\ 0 & 0 & 0 & -1 & 0 \\ 0 & 0 & 0 & 0 & 1 \end{pmatrix}, \qquad M = \begin{pmatrix} m & 0 & 0 & 0 & 0 \\ 0 & m & 0 & 0 & 0 \\ 0 & 0 & m & 0 & 0 \\ 0 & 0 & 0 & m & 0 \\ 0 & 0 & 0 & 0 & m \end{pmatrix} .$$

By analogy with the Dirac equation we also define the conjugate field (a row-vector, whereas u denotes a column vector):

$$\bar{u} = u^\dagger \beta^4 = (u_0, -u_1, -u_2, -u_3, u_4), \; u^\dagger = u^{*T} \qquad (2.49)$$

The first-order system of equations satisfied by u, $\bar{u}$ has the form (Duffin-Kemmer-Petiau equations [1])

$$(i\beta^k \partial_k - M)u = 0, \quad i\partial_k \bar{u}\beta^k + \bar{u}M = 0, \; k = 0, 1, 2, 3. \qquad (2.50)$$

These equations are the Euler equations of the action

$$S_u = \int \{\tfrac{i}{2}[\bar{u}\beta^k \partial_k u - (\partial_k \bar{u})\beta^k u] - \bar{u}Mu\} d^4x, \qquad (2.51)$$

which differs by a factor of 1/m from the action (2.43).

The action is invariant under the group U(1), i. e., if the fields u, $\bar{u}$ are subjected to the gauge transformations of the first kind:

$$u \to \exp(ie\chi)u, \quad \bar{u} \to \bar{u}\exp(-ie\chi), \qquad (2.52)$$

where χ is a number, or operator, independent of x. Therefore, by the Noether theorem the "current"

$$j^k = e\bar{u}\beta^k u, \qquad (2.53)$$

which differs from the current (2.45) by a factor of m, is conserved:

$$\partial_k j^k = 0. \qquad (2.54)$$

Since the gauge transformation (2.52) leaves all the observable quantities, including current and 4-momentum, invariant, nothing prevents different observers at different points (separated by spacelike distances, so that they cannot influence each other) from choosing different "gauges" χ, i. e., replacing the constant χ in (2.52) by

[1] The Duffin-Kemmer matrices are characterized by the following trilinear commutation property: $\beta^k\beta^n\beta^m + \beta^m\beta^n\beta^k = \eta^{kn}\beta^m + \eta^{mn}\beta^k$.

a smooth function of the point x in Minkowski space, which we also denote by $\chi(x)$ (we have factored out the "electric charge" e, but this is not necessary); this leads to the local gauge transformation

$$u \to \exp(ie\chi(x))u, \quad \bar{u} \to \bar{u}\exp(-ie\chi(x)) \tag{2.55}$$

The transformation (2.55) can be viewed as consisting of a copy of the group U(1) attached to each space-time point x, and acting on the two-component field $u(x)$, $\bar{u}(x)$ (we count here only the "internal" components; one should not forget that u is a 5-component vector in the Duffin-Kemmer index; the transformation for $\bar{u}$ has been written with the group action from the right in order to accomodate operator gauge transformations, where the gauge function $\chi(x)$ will be an operator). In this case one can consider the exponent $\pm ie\chi(x)$ as an element of the Lie algebra $i\mathbb{R}$ of the group U(1).

The action (2.51) is no longer invariant under the gauge transformation (2.55). Indeed, with $g(x) = \exp[ie\chi(x)]$, $g^{-1}(x) = \exp[-ie\chi(x)] = g(x)^*$, the variation of S_u becomes:

$$\delta S_u = \int \frac{i}{2}[\bar{u}g^{-1}(x)\beta^k\partial_k g(x)u - (\partial_k\bar{u}g^{-1}(x))\beta^k g(x)u]d^4x$$

$$- \int \frac{i}{2}[\bar{u}\beta^k\partial_k u - (\partial_k\bar{u})\beta^k u]d^4x = -\int e\bar{u}\beta^k u\partial_k\chi d^4x = -\int j^k\partial_k\chi d^4x. \tag{2.56}$$

The presence of this term can be compensated by adding to the action S_u

$$S_{uA} = -\int j^k A_k(x)d^4x, \tag{2.57}$$

where $A_k(x)$ is the covariant vector which can be considered the coefficient of a 1-form A, which under the gauge transformation (2.55) is subject to the gauge transformation of the second kind ($g^{-1}dg = ied\chi$)

$$A \to A - \frac{i}{e}g^{-1}dg, \quad A_k \to A_k + \partial_k\chi. \tag{2.58}$$

Since A has been combined with $d\chi$ and the latter is to be considered a hermitian element in the Lie algebra of U(1), we must consider A itself to be a Lie-algebra valued 1-form (in this case, where the Lie algebra is isomorphic to the real line, A is an ordinary real 1-form, but this viewpoint will become important when we discuss a nonabelian gauge group). For Lie-algebra-valued differential forms, cf. 1.4.

Another way of looking at this problem is to note that the Duffin-Kemmer-Petiau equations (2.50) are not invariant under the local gauge transformation (2.55), but that the equations can be modified by replacing the ordinary partial derivatives ∂_k (the gradient) by the "gauge-covariant derivative" obtained from it by adding (or subtracting) the 1-form ieA_k:

$$\nabla_k = \partial_k \pm ieA_k, \qquad (2.59)$$

or, equivalently, replacing the exterior differential d by the covariant exterior differential

$$Du = du + ieAu, \quad D\bar{u} = d\bar{u} - ieA\bar{u}. \qquad (2.60)$$

In (2.59) the upper sign corresponds to the differentiation of u, and the lower sign corresponds to the differentiation of $\bar{u}$.

Since we are dealing with an abelian group and an abelian Lie algebra, the exterior covariant differential of A coincides with the ordinary exterior differential (this can be seen easily in terms of components:

$$DA = (\partial_i A_k \pm ieA_i A_k) dx^i \wedge dx^k = \tfrac{1}{2}(\partial_i A_k - \partial_k A_i) dx^i \wedge dx^k = dA, \quad (2.61)$$

where the symmetric term $A_i A_k$ cancels on contraction with the antisymmetric $dx^i \wedge dx^k$; this can also be interpreted by saying that the new field A has charge $e = 0$). We are thus led to the electromagnetic field 2-form

$$F = dA = DA, \quad F_{ik} = \partial_i A_k - \partial_k A_i, \qquad (2.62)$$

which automatically satisfies the Bianchi identity

$$DF = dF = 0, \qquad (2.63)$$

equivalent to the homogeneous pair of Maxwell equations, as we have seen in Section 2.1.

In order to obtain the second pair of Maxwell equations, we have to add to the action of the u-field (in which, in order to take into account the term S_{uA}, we can replace the partial derivatives by the covariant derivatives (2.59)) the action of the 1-form A, which as we have seen in Section (2.2) is of the form (2.41).

This leads to the equation

$$D{*}F = d{*}F = {*}J, \tag{2.64}$$

where $*J$ is the 3-form dual to the current j^k:

$$*J = e_{ikmn}(e\bar{u}\beta^i u)\,dx^k \wedge dx^m \wedge dx^n/3! . \tag{2.65}$$

The reason we have written both D and d in Eqs. (2.62) - (2.64) is the fact that we will have to generalize these equations to the nonabelian case.

We shall return later to the problems related to the quantization of the coupled field u, $\bar{u}$, A, and now turn our attention to the simplest generalization of the preceding reasoning to a nonabelian group.

2.3.1. The Yang-Mills Gauge Theory. We now consider the case of a scalar field u, $\bar{u}$ with an "internal" degree of freedom described by the group SU(2). Historically, Yang and Mills [63] treated in 1954 the analogous case of a spinor field, but since we are interested only in the "internal" degrees of freedom, we might as well stick to the now familiar spinless fields.

The internal label will be denoted by a, and will take on the values 1, 2. Thus the field u will now carry the additional index a, and so will the conjugate field $\bar{u}$: u_a, $\bar{u}_a$. In the complex two-dimensional vector space we have a hermitian inner product $\langle u,v\rangle = \sum_a \bar{u}_a v_a$, which is left invariant under unitary transformations, in particular under the group SU(2) (we do not mention explicitly the other inner product, in 5-dimensional space, which will always be understood).

The field equations

$$i\beta^k \partial_k u_a - M u_a = 0, \quad a = 1, 2$$
$$i\partial_k \bar{u}_a \beta^k + \bar{u}_a M = 0, \tag{2.66}$$

will be invariant under the action of SU(2), if it is understood that if u_a is acted upon by a matrix $g \in SU(2)$ then $\bar{u}_a$ is acted upon by the complex-conjugate matrix $\bar{g}$, acting on the right, i. e., by $g^+ = \bar{g}^T = g^{-1}$. One may state this differently, saying that the matrices β^k, M, commute with the representation of SU(2) under which u, $\bar{u}$ transform.

Noether's theorem tells us that in this case there are three conserved currents j_α^k, $\alpha = 1, 2, 3$, corresponding to the three parameters of the isospin group SU(2):

$$j_\alpha^k = \sum_{ab}\sum i\bar{u}_a(\tau_\alpha)_{ab}\beta^k u_b, \qquad (2.67)$$

where the matrices τ_α ($\alpha = 1, 2, 3$) are the generators of the Lie algebra of the group SU(2), i. e., the isospin Pauli matrices. Here we have absorbed a possible "charge" into the fields to which the current (2.67) will be coupled. The appropriate "total charges"

$$Q_\alpha = \int j_\alpha^0 d^3x, \qquad (2.68)$$

are a representation of the Lie algebra $su(2)$ of the gauge group, and hence must satisfy the commutation relations

$$[Q_\alpha, Q_\beta] = \sum_\gamma ie_{\alpha\beta\gamma}Q_\gamma, \qquad (2.69)$$

pointing to the fact that we are dealing in fact with a quantum theory of the fields u, $\bar{u}$.

We can repeat Weyl's heuristic argument which led us to the potential 1-form in the case of the abelian group U(1), with the following result. In place of the constant SU(2) transformations we require that the fields u_a be subject to a "point-dependent gauge transformation"

$$u_a \to g(x)u_a \ , \quad g(x) \in SU(2), \qquad (2.70)$$

where g(x) is a smooth function on Minkowski space with values in SU(2). In other words, we subject u (and $\bar{u}$) to the action of a matrix-valued function on Minkowski space, or as we shall see in the next chapter, the action of a <u>section</u> of the trivial bundle $M \times SU(2)$. In this the action of the field u, which coincides with (2.51), with the ordinary product $\bar{u}u$ replaced by the inner product $\langle u,u \rangle$ acquires an additional term analogous to (2.56) (for simplicity we write only one of the terms in the lagrangian density; the other term leads to a factor of 2 in the result, which cancels against the 2 in the denominator). Thus, for the first term in the integrand of (2.51)

$$L_1 = \frac{i}{2}\sum\bar{u}_a\beta^k\partial_k u_a \qquad (2.71)$$

we have the following expression of the change produced by the action of g(x) (for simplicity, the sums are suppressed, summation is over all pairs of repeated indices, wherever they occur):

$$\begin{aligned}\delta L_1 &= (i/2)[(\overline{g_{ab}(x)u_b})\beta^k\partial_k(g_{ac}(x)u_c) - \bar{u}_a\beta^k\partial_k u_a] \\ &= (i/2)[\bar{u}_b({}^Tg^*)_{ba}\beta^k(g_{ac}(x)\partial_k u_c + (\partial_k g_{ac}(x))u_c) - \bar{u}_a\beta^k\partial_k u_a] \\ &= (i/2)(g^{-1}(x)\partial_k g(x))_{bc}\bar{u}_b\beta^k u_c, \qquad (2.72)\end{aligned}$$

where we have used the fact that g(x) is a unitary matrix: ${}^Tg^* = g^+ = g^{-1}$ and cancelled the identical terms. A similar term is contributed by the second term of the integrand of (2.51), if one uses the identity $g^{-1}dg = -dg^{-1}g$, thus leading to

$$\delta S_u = \int i(g^{-1}(x)\partial_k g(x))_{bc}\bar{u}_b\beta^k u_c d^4x. \qquad (2.73)$$

The expression

$$g^{-1}(x)dg(x) = g^{-1}\partial_k g\, dx^k = \sum_1^3 \omega_\alpha\tau_\alpha \qquad (2.74)$$

is the Maurer-Cartan 1-form of the group G, i. e., a Lie-algebra valued 1-form, which can be expanded in terms of the three generators τ_α (α = 1, 2, 3) of SU(2) (the τ_α are the isospin Pauli matrices), which are usually incorporated in the three "isospin currents"

$$j_\alpha^k = \bar{u}_a(\tau_\alpha)_{ab}\beta^k u_b, \qquad (2.75)$$

and the ω_α are scalar-valued 1-forms (the differentials of the three gauge parameters of SU(2)).

In the same manner as for the abelian case, the variation (2.73) can be compensated by introducing three 1-forms T^α, or the Lie-algebra valued one-form

$$Y = \sum T^\alpha\tau_\alpha, \quad T^\alpha = A_k^\alpha dx^k, \qquad (2.76)$$

the Yang-Mills potential one-form (the usual Yang-Mills potentials are the covariant vector fields A_k^α), coupled to the currents j_α^k.

Alternatively, in complete analogy with the abelian case, we can replace the ordinary differentiation of the fields u_a by the "covariant differential"

$$Du = du + iYu = (\partial_k u + i\sum A_k^\alpha\tau_\alpha u)dx^k . \qquad (2.77)$$

In distinction from the case of the abelian gauge group U(1), the potential one-form Y will undergo a transformation generalizing (2.58) where in addition to the Maurer-Cartan one-form $g^{-1}(x)dg(x)$ the matrix-valued one-form is subjected to the appropriate transformation of the adjoint representation of G:

$$Y \to \text{Ad}(g^{-1})Y + g^{-1}(x)dg(x), \qquad (2.78)$$

or, in terms of the Yang-Mills potentials A_k^α

$$A_k^\alpha \tau_\alpha \to g^{-1}(x)A_k^\alpha \tau_\alpha g(x) + g^{-1}(x)\partial_k g(x). \qquad (2.79)$$

in which the reader will recognize the gauge transformations of the second kind introduced by Yang and Mills (we shall return later to a more detailed discussion of these transformations).

The Yang-Mills field-strength two-form (which is the analogue of the electromagnetic field-strength two-form F) is obtained by taking the <u>covariant differential</u> of the one-form Y (we denote this two-form by M, so that the complete Yang-Mills field is denoted by Y-M):

$$M = DY = dY + Y\wedge Y = dY + \tfrac{1}{2}[Y, Y]. \qquad (2.80)$$

Just like Y, M is a Lie-algebra-valued two-form. The last form of this equation is analogous to the Maurer-Cartan structure equation (1.57). The expression for the Yang-Mills field strength in terms of coordinates is

$$M = \sum \tau_\alpha M_{jk}^\alpha dx^j\, dx^k; \quad M_{jk} = \partial_j Y_k - \partial_k Y_j + \tfrac{1}{2}\sum c_{\beta\gamma}^\alpha Y_j^\beta Y_k^\gamma, \qquad (2.81)$$

where the structure constants of the group SU(2) are the components of the completely antisymmetric tensor of rank three.

Taking the covariant exterior differential of Eq. (2.80) we obtain the first Yang-Mills equation as a Bianchi identity for the two-form M:

$$DM = DDY = 0. \qquad (2.82)$$

We leave it to the reader to write this equation out in coordinates.

At this point we note that Eqs. (2.80) and (2.82) are characteristic for connections and their curvature forms, objects which we will study in detail in the next chapter.

By analogy with electromagnetism, we introduce now the dual form

$*M = eM$, where e is the fourth-rank antisymmetric tensor, and the dual three-form $*J$, dual to the current four-vector j, introduced above. Then the second Yang-Mills equation is the analog of the inhomogeneous Maxwell equation:

$$D*M = *J. \tag{2.83}$$

If one wants to derive these equations from an action principle one has to use one of the two four-forms one can build up out of M and $*M$, to be integrated over the four-volume. The two candidates for the Yang-Mills action are then

$$\|M\|^2 = \tfrac{1}{4}\int M \wedge *M \qquad \text{or} \qquad \tfrac{1}{4}\int M \wedge M; \tag{2.84}$$

we shall see that the first will be the choice of the action, whereas the second, appropriately normalized, will play an important role in characterizing classical solutions of the Yang-Mills equations by the appropriate "characteristic class", yielding an integer for that integral. We leave it as an exercise for the reader to translate all the expressions of this section into coordinate language and to compare them with the usual expressions. The economy in writing will then be obvious.

2.4. Cohomology of the Electromagnetic Field and Magnetic Monopoles. In this section we sketch briefly some aspects of the description of the electromagnetic field in terms of differential forms which could be called "de Rham cohomology of the electromagnetic potential". The main ideas of this approach have been described to the author verbally by F. Strocchi, but the contents of this section may not be identical to Strocchi's results as they will ultimately be published. Similar and related ideas have been proposed by J. Roberts [45] and V. Lugo.[34] For a comprehensive review of the literature on magnetic monopoles, cf. a forthcoming review paper by R. Brandt and J. Primack (to be published in Reviews of Modern Physics).

We start from the observation that the electromagnetic field two-form F is closed in the absence of magnetic monopoles, $dF = 0$, $F = dA$.

One is tempted to consider the description in terms of the potential one-form A redundant, though convenient, as long as one remains in the framework of classical theory. However, as soon as the electromagnetic field is coupled in a gauge-invariant way to a quantized object, e. g., the wave function describing an electron, the Bohm-Aharonov effect shows that the potential (or more correctly, some invariants associated to it) can manifest itself through interference effects. In fact, the phase factor $\exp(i\oint_\gamma A)$, where we have as usual absorbed the charge into A, and set Planck's constant and the speed of light equal to one, produces a fringe shift in the two beams surrounding a region of nonzero A; this phase factor can be transformed, by means of Stokes' theorem into the factor $\exp(i\int_\sigma F)$ where the integral is over the surface σ bounded by the loop γ, i. e., depends on the magnetic flux. One is thus clearly faced with topological (homological and cohomological) problems when discussing electromagnetism coupled to quantized objects.

If one assumes that there is also a magnetic charge (or current), i. e., that F is not closed, but

$$dF = J_M, \tag{2.85}$$

then one is forced to the conclusion that there cannot exist a smooth 1-form A, and a careful analysis of this situation carried out by Yang and Wu [62] (cf. also [34]) has shown that the Dirac relation between electric and magnetic charge can be derived simply from the requirement that the gauge transformation be single-valued (since the gauge group is U(1) a Maurer-Cartan connection has the form du/u, where u is any smooth complex-valued function of x, having absolute value one).

Thus, regions carrying magnetic charge would appear as holes in the underlying manifold, and the fact that outside these holes F is closed, and its integral is an integer (when normalized appropriately) will lead to a cohomological characterization of magnetic charge.

Returning to the case of ordinary electromagnetism, without magnetic charge, we observe that similar cohomological questions would arise when one attempts to interpret the second pair of Maxwell equations:

$$d*F = *J, \qquad d*J = 0, \tag{2.86}$$

as equations for operator-valued distributions. This problem is being invetigated by Strocchi [50] and in a different light, by Roberts[45], and will play an important role in the quantum-field theory context.

Here we only mention those aspects which may already be important on a classical level. In particular, in the quantum-field theory context, if one wishes to define the fields as operator-valued distributions, it is important to have at one's disposal classes of test-functions which are solutions of the classical field equations. As is well known, locality properties, related to the support of test functions are quite important. In this connection, we consider here two-forms on Minkowski space which belong to the de Rham spaces $\mathcal{D}$ or $\mathcal{S}$ of functions with compact support or of rapid decrease, which are infinitely differentiable. This means, that the coefficients of these forms have the appropriate properties. Strocchi has proved that such forms admit a local Hodge decomposition.

A two-form F is said to admit a Hodge decomposition if it can be written in the form

$$F = F_1 + F_2, \tag{2.87}$$

with

$$dF_1 = 0 \quad \text{and} \quad d*F_2 = 0. \tag{2.88}$$

A Hodge decomposition is called local when the supports of the forms F_1 and F_2 are in a convex open set containing the support of F. It is easy to show that if a form F has compact support and satisfies $dF = d*F = 0$, then $F = 0$ (since then $-(d d* + d*d)F = \Box F = 0$, and $F = 0$ outside a compact set).

Many other properties of charges and monopoles can be translated into cohomological language. We will treat some nonabelian analogs later in these notes, and will return to a more detailed discussion elsewhere.

3. PRINCIPAL FIBRATIONS AND ASSOCIATED VECTOR BUNDLES

3.0. Introduction and Motivation. This chapter is the core of the differential-geometric approach to gauge theories. In it we develop the fundamental notions of principal fibration (principal fiber bundle) and associated vector bundle. These concepts, which have been introduced into differential geometry and topology in the fifties, are the natural framework in which to discuss gauge theories, and have in fact been implicit in much of the physical literature.

In Section 1.1.8 we have introduced the tangent bundle to a manifold as the collection of all tangent spaces. This is the prototype of all vector bundles to be discussed below: roughly speaking, a vector bundle is a family of vector spaces, each attached to a point of a manifold, such that locally (i. e., in a neighborhood of a point) the bundle looks like the product of that neighborhood with the given vector space (called the fiber). Globally, as we have seen for the tangent bundle to the sphere in 3-space, the bundle is no longer a product, unless the bundle is "trivial".

In Section 2.3 we encountered a gauge group, which is for us the prototype of another bundle: a principal fibration or principal fiber bundle. In this case we attached to each spacetime point a different copy of a given Lie group (which acted on a vector space attached to each point, i. e., acted on the fibers of a vector bundle). We can think of a principal fibration as the result of glueing together a collection of products of open sets of a manifold and copies of a given Lie group, such that, as we move from point to point, the group undergoes a "twist".

A familiar example of a vector bundle which is not trivial, i. e., which is not isomorphic to the product of a manifold and a vector space, is the Möbius strip. Here the manifold over which the fibers "sit" is the circle S^1, and the fiber is a one-dimensional vector space (or the unit interval [0, 1]), with the fibers given a twist as one goes around the base circle, such that after a full circumference the opposite ends of the interval are identified. The corresponding trivial bundle is the cylinder, where the fiber is not twisted at all, or a strip with a double twist of 2π, which is topologically equivalent.

In the vector bundles and principal fibrations which appear in gauge theories, the bundles are usually trivial in the mathematical sense. However, when sources or singularities are present, since we require smoothness of the various functions, vector fields and connections (see below) involved, the base spaces will cease to be simply connected, and thus the topological structure of these manifolds becomes important, and similar to the situation in electromagnetism discussed in Section 2.4, cohomological properties become important, and the appropriate gauge fields will be assigned to various characteristic classes (Chern classes in the complex case, Pontryagin classes in the case of real vector spaces and orthogonal groups) with associated integers, which have played an important role in recent developments. These aspects will be discussed in a later chapter.

In our discussion of the electromagnetic field and of the Yang-Mills field we have seen that two concepts forced themselves upon us. The first is the concept of a <u>section</u> of the principal fibration, i. e., the choice of a group element at each point of the underlying manifold (spacetime), varying smoothly (i. e., in an infinitely differentiable manner) from point to point. This was exactly the choice of a gauge. It will be shown in this chapter that the existence of a global section (i. e., the possibility of choosing a global smooth gauge function throughout the whole of spacetime) is equivalent to the triviality of the fibration. This would settle the problem for the mathematician, but for the physicist two different trivializations of the same fibration may have different significances; indeed, anyone who has done a calculation in quantum electrodynamics will be aware of the important role played by an appropriate choice of gauge. This is even more so in the case of gauge theories with nonabelian gauge groups (cf., e. g., the detailed discussion of these aspects in the review by Abers and Lee [1]).

The other concept which our preliminary discussion brought to light was the fact that the vector potential and the field strength are to be viewed respectively as a one-form and a two-form with values in the Lie algebra of the gauge group, the one-form being related to the Maurer-Cartan form of the Lie group. We saw that one way of introducing the gauge field was through the "gauge-covariant derivative". These aspects of the gauge fields will lead us to identify the gauge potential with a connection in the principal fibration with the gauge group as structure group and the gauge field will be identified with the curvature two-form of the connection.

The particle fields which undergo gauge transformations will be identified with sections of vector bundles associated to the principal fibration through various representations of the gauge group, i. e., with a choice of a vector at each point of the underlying spacetime, varying smoothly from point to point. The existence of a connection in the principal fibration makes it possible to split the local vector spaces (fibers) of the vector bundle into a "horizontal" and a "vertical" subspace, thus leading to the notion of "parallel transport" of vectors, and hence to the idea of covariant derivative. These concepts will be developed in detail in the next chapter, where we will also introduce the idea of holonomy (i. e., the effect of parallel transport along closed loops in the base space).

Some of the physical implications will be mentioned in passing throughout these chapters, but a complete discussion of the physics is concentrated in the following chapters.

Finally, a brief remark about notations. For typographical reasons I have tried to avoid, as far as possible the use of Greek letters.

3.1. Fibrations. We follow the terminology of Bourbaki and Dieudonné which differs slightly from that of other authors, which use the term fiber space, or fiber bundle for the same object. In this chapter, as in Chapter 1, we consider only smooth fibrations, i. e., all manifolds and maps between them will be considered C^∞, and this fact will not always be stated explicitly. One can rewrite most of the definitions assuming differentiability of order r, but at the cost of having to watch out for possible loss of differentiability; some concepts can be defined for topological spaces only. The reader should be warned when looking into the literature to watch out whether a result he uses is valid in the context in which he needs it (which is often the case).

3.1.1. Definition. A smooth fibration (or simply fibration) is a triple $\lambda = (X, B, p)$, where the base space B and the bundle space (or total space) X are smooth manifolds and p is a C^∞-mapping of X into B which is surjective and satisfies the condition of local triviality:

(F) For any $x \in B$ there exists an open neighborhood U of x, a manifold F and an isomorphism (diffeomorphism) φ of $p^{-1}(U)$ onto $U \times F$ such that $p(\varphi^{-1}(x, y)) = x$, for all $x \in U$ and $y \in F$.

The mapping p is called the projection of the fibration and for each $b \in B$ the inverse image $X_b = p^{-1}(B)$ is a closed submanifold of X, called the fiber over b. The local triviality condition implies that all $X_{b'}$ are diffeomorphic to X_b for all $b' \in U$, a neighborhood of b. Therefore one can picture the fibration a a bundle of fibers or fiber bundle, particularly, when all the fibers are diffeomorphic to the same manifold F (called the typical fiber). We then say that the fibration is a bundle of fiber-type F. Thus, if F is a Lie group we will be led to a principal fibration, if F is a fixed vector space we will have a vector bundle, etc.

The tangent space at a point x X to the fiber $X_{p(x)}$ will be identified to a subspace of $T_x(X)$ and the vectors of that subspace will be called vertical tangent vectors.

3.1.2. Examples:

(i) The trivial fibration or trivial bundle over B with fiber-type F is the product-triple $(B \times F, B, pr_1)$, where pr_1 denotes projection on the first factor of the product $pr_1(b, f) = b$, $b \in B$, $f \in F$. Here B and F are manifolds of an arbitrary nature.

(ii) Let M be a manifold and TM its tangent bundle; then the triple (TM, M, p), where p is the C^∞-map that associates to each tangent space of the manifold the point of M at which the tangent space is taken is a fibration which we can consider the prototype of all vector bundles to be discussed later. We recall that if M has dimension n, TM has dimension 2n, and the subspace of vertical vectors has dimension n.

(iii) Similarly, T*M, the cotangent bundle of a manifold, and the various tensor bundles are fibrations in a natural manner.

(iv) The Möbius strip discussed in the introduction is a fibration with base-space S^1, the unit circle, and typical fiber the segment [0,1].

(v) If $\lambda = (X, B, p)$ and $\lambda' = (X', B', p')$ are twc fibrations the product of these fibrations is defined as

$$\lambda \times \lambda = (X \times X', B \times B', p \times p').$$

(vi) A fibration with base space B whose fibers are discrete is called a covering space of B. Well-known examples are the Riemann surfaces of the square root or the logarithm.

3.1.3. Morphisms of fibrations. A morphism of the fibration $\lambda = (X, B, p)$ into the fibration $\lambda' = (X', B', p')$ is a pair of C^∞-mappings (f, g), $f: B \to B'$, $g: X \to X'$, such that $p' \circ g = f \circ p$.

An isomorphism is a morphism where f and g are diffeomorphisms, its inverse is the pair (f^{-1}, g^{-1}).

When $B = B'$ and (Id_B, g) is a morphism we call g a B-morphism of X into X'.

A fibration $\lambda = (X, B, p)$ is trivializable if there exists a B-isomorphism of λ onto a trivial fibration $(B \times F, B, pr_1)$ and the isomorphism is called a trivialization. For the physical applications, where many fibrations will be trivializable, it is important to note that not all trivializations are equally useful. If f_1, f_2 are two trivializations of the fibration λ onto the trivial bundle $B \times F$, then $f_2(x) = h(g_1(x))$, where $h: (b, f) \mapsto (b, h(b,f))$ is a B-automorphism of the trivial bundle $B \times F$, i. e., a C^∞-mapping such that for each $b \in B$ $h(b, \cdot)$ is a diffeomorphism of F onto itself.

If in the definition of a morphism of fibrations $B = B'$ and $f = Id_B$, we say that g is a B-morphism of λ into λ'; if g is an isomorphism and g^{-1} is a B-morphism, then g is a B-isomorphism of λ onto λ'.

3.2. Sections. A section of the fibration $\lambda = (X, B, p)$ is a morphism (C^∞-mapping) $s:B \to X$ such that $p \circ s = Id_B$. A section can be viewed as a B-morphism of the fibration (B, B, Id_B) into (X, B, p). Sometimes the condition of infinite differentiability of sections is replaced by a weaker condition, such as continuity or differentiability of order r. We will always understand by section a C^∞- section defined globally, i. e., on all B.

Any trivializable fibration admits at least one section, but the converse is not generally true; thus, vector bundles (cf. infra) always admit the zero section, whether they are trivializable or not, but principal bundles admitting a section are always trivializable.

A section in a trivial bundle $(B \times F, B, pr_1)$ is a mapping $b \mapsto (b, f(b))$, where f is a C^∞-mapping from B into F; i. e., there is a bijective correspondence between sections and graphs of C^∞-functions from B into F, and one is often tempted to use this identification in nontrivial bundles also. In the latter case this correspondence is only local, and should not be taken too literally.

If $g: X \to X'$ is a B-morphism, each section s of (X, B, p) becomes a section $g \circ s$ of (X', B, p'). If $(f, g): (X, B, p) \to (X', B', p')$ it is not generally possible to define the image of a section in (X, B, p), unless f is a diffeomorphism of B onto B': then $s':b' \mapsto g(s(f^{-1}(b')))$.

3.3. Inverse images (pullbacks) and fibered products. This section may be omitted on a first reading and used only for reference as the need arises. For details of the proofs the reader is referred to Dieudonné, Vol. III, p.81-83.

Let $\lambda = (X, B, p)$ be a fibration and B' be a manifold mapped into B by the C^∞-map f. The the following theorem, the proof of which can be found in Dieudonné, defines the inverse image $f^*(\lambda)$ of λ under f:

(i) The set $B' \times_B X$ of points $(b', x) \in B' \times X$ such that $f(b') = p(x)$ is a closed submanifold of $B' \times X$, called the fibered product of B' and X over B.

(ii) If one denotes by p' the restriction to $B' \times_B X$ of the projection pr_1 on the first factor in $B' \times X$, then $\lambda' = (B' \times_B X, B', p')$ is a fibration such that the fiber $(B' \times_B X)_{b'}$ over each point b' is canonically diffeomorphic to $X_{f(b')}$, the fiber of λ over $f(b')$. Denoting by f' the restriction of pr_2 to $B' \times_B X$, the pair (f, f') is a morphism of λ' into λ. The fibration λ' is called the inverse image or pullback of λ under f and is denoted by $f^*(\lambda)$.

If λ is trivial λ' is also trivial. For each section $s: B \to X$ of λ the mapping $s': b' \to (b', s(f(b')))$ is a section of $f^*(\lambda)$ called the pullback (inverse image) of s under f and will be denoted by $f^*(s)$. It will usually be clear from the context whether we are dealing with the pullback of a section of a fibration or the pullback of a differential form, for which we have also used the notation f^*.

These concepts can be adapted to the case when B' is a submanifold of B and j is the canonical injection. In that case $j^*(\lambda)$ is the fibration induced on B' by λ and $j^*(s)$ is the restriction of the section s to B'. The set of sections of λ over B' is denoted by $\Gamma(B', X)$; these should be distinguished from the global sections (over B), and may exist even when the latter do not exist.

The condition (F) in 3.1 can be restated by saying that each $b \in B$ has a neighborhood U such that the induced fibration is trivializable.

The <u>fibered product</u> of two fibrations $\lambda = (X, B, p)$ and $\lambda' = (X', B, p')$ is denoted by $\lambda \times_B \lambda' = (X \times_B X, B, p \times_B p')$ and defined as follows: consider the product fibration

$$\lambda \times \lambda' = (X \times X', B \times B, p \times p'),$$

(the two base manifolds are identical) and let $\delta: B \to B \times B$ denote the diagonal map (i. e., the map associating to the point b in B the pair (b, b) in $B \times B$). Let $\lambda'' = \delta^*(\lambda \times \lambda')$ be the pullback of the product by δ. The total space of this fibration is a submanifold of $B \times X \times X'$ consisting of those triples (b, x, x') for which $b = p(x) = p'(x')$. The set $X \times_B X'$ such that $p(x) = p'(x')$ is a submanifold of $X \times X'$ (the fibered product of the two manifolds). So up to a symmetry interchanging X and X' the space X" of the fibration λ'' is the graph of the restriction of $p \circ pr_1$, or $p' \circ pr_2$ to the submanifold $X \times_B X'$ of $X \times X'$, allowing one to identify these two submanifolds (canonically). If p" denotes the projection of the fibration λ'' and $U \subset B$ is a subset where X and X' are simultaneously trivializable, there are diffeomorphisms $\varphi: p^{-1}(U) \to U \times F$, $\varphi': p'^{-1}(U) \to U \times F'$, such that

$$(b, (y, y')) \to (b, \varphi^{-1}(b, y), \varphi'^{-1}(b, y')), y \in F, y' \in F'$$

is a diffeomorphism of $U \times (F \times F')$ onto $p''^{-1}(U)$, i. e., a trivialization of the restriction of λ'' to U. The fiber X''_b is diffeomorphic to $X_b \times X'_b$. The fibration λ'' is called the fibered product of λ and λ' over B and is denoted by $\lambda \times_B \lambda'$ (or by abuse of notation, by $X \times_B X'$).

Finally, we quote another result, the proof of which can be found in Dieudonné, Vol. III, (16.12.12):

If the fibers of the fibration $\lambda = (X, B, p)$ are diffeomorphic to $\mathbb{R}^n$ there exists a global section of λ over B.

This applies, in particular, to vector bundles, which, as we have already mentioned, admit the zero-section as a global section.

<u>3.4. Definition of fibrations by means of charts.</u> In the same manner in which we defined a manifold in terms of charts, one can define fibrations by means of charts.

Since the base space of a fibration is a manifold, there always exists a covering (U_i) of B defining the charts of an atlas of the manifold B. On the other hand, by definition, the covering (U_i) can be chosen in such a way that the restriction of the fibration to each open set U_i is trivializable, i. e., there exists a diffeomorphism φ_i of the set $p^{-1}(U_i)$ onto the product $U_i \times F_i$, where F_i is the fiber at some point x in U_i.

$$\varphi_i\colon p^{-1}(U_i) \to U_i \times F_i. \tag{3.1}$$

Now consider two sets U_i and U_j of the covering with nontrivial intersection $U_i \cap U_j$ and denote by φ_{ij} the restriction of φ_i to the set $p^{-1}(U_i \cap U_j)$:

$$\varphi_{ij}\colon p^{-1}(U_i \cap U_j) \to (U_i \cap U_j) \times F_i. \tag{3.2}$$

Since on $U_i \cap U_j$ we have two trivializations (the first one given by (3.2), the second one obtained by interchanging i and j, i. e., restricting the diffeomorphism $\varphi_j\colon p^{-1}(U_j) \to U_j \times F_j$ to $p^{-1}(U_i \cap U_j)$), the composite diffeomorphism

$$\psi_{ji} = \varphi_{ij} \circ \varphi_{ji}^{-1}\colon (U_i \cap U_j) \times F_i \to (U_i \cap U_j) \times F_j, \tag{3.3}$$

is of the form

$$(b, y) \mapsto (b, f_{ji}(b, y)), \quad b \in B, \; y \in F_i,$$

where f_{ji} is a C^∞-mapping into F_j (one can think of it as a mapping from F_i to F_j, depending on the point; in the case of vector bundles this will be a linear mapping, and for principal fibrations it will be a group isomorphism, cf. infra). The diffeomorphism ψ_{ji} is sometimes called the <u>transition function</u> between the two bundle charts.

If we now consider three sets of the open covering, U_i, U_j, U_k, and restrict the transition functions to the intersection $U_i \cap U_j \cap U_k$. Then these transition functions satisfy the identity ("patching")

$$\psi_{ki} = \psi_{kj} \circ \psi_{ji}. \tag{3.4}$$

Conversely, given a manifold B, with an open covering (U_i) and for each index i a manifold F_i (the fiber), such that for each pair of indices we have a transition function (3.3) satisfying (3.4), then one right and $\pi : P \to B$ is a C^{∞}-surjection of P onto B (a morphism of manifolds), the whole structure being subject to the condition of <u>local triviality</u>:

(P) For each $b \in B$ there exists an open neighborhood U of b and an isomorphism $f: U \times G \to \pi^{-1}(U)$, such that for any point u U and any group elements $g, g' \in G$ we have

$$(f(u,g)) = u \quad \text{and } f(u, gg') = f(u, g)\cdot g', \tag{3.5}$$

where $x \cdot g$ denotes the right action of the group on elements x of P.

The triplet (P, B, π) is a fibration with projection π, and the equivalence relation induced in P by the projection is the same as the one defined by the right action of the group G. In other words, the fibers of the fibration (P, B, π) coincide with the orbits of the right action of G on P, or the points projected onto the same point b of the base are transformed into one another by the action of the group G on the manifold P. It is convenient to picture the base as a "horizontal" set; then each fiber is "vertical" and the group action is also vertical. We will often use this terminology.

<u>3.5.3. Examples.</u> i) Let X be a manifold and let G be a Lie group acting freely and differentiably on the right. Suppose the orbit manifold X/G exists and let $\pi: X \to X/G$ be the canonical C^{∞}- surjection which associates to any point of x its orbit under G. Then it is not hard to see that the triplet $(X, X/G, \pi)$ is a fibration, and since the action of G takes the fibers $\pi^{-1}(b)$ into themselves, it is a principal fibration. In fact, every principal fibration may be thought of as a fibration of this type.

ii) Let H be a Lie group and G a Lie subgroup acting on H to the right. Then the coset space $B = H/G$ is a homogeneous space which is the base of the principal fibration $(H, H/G, G, \pi)$, where π is the canonical projection on the orbits.

iii) A gauge group, as described in Section 2.3, is a principal fibration with base space the Minkowski space (with possible points can define the topological space X (the space of the bundle) by patching together the sets $U_i \times F$ by means of the homeomorphisms ψ_{ji}. This can be shown to lead to a metrizable, separable and locally compact space X, which in fact is a differentiable manifold (its charts are the bundle charts $U_i \times F$). The projection map $p: X \to B$ is defined in each chart by projection onto the first component of the product, and thus one can think of the fibration (X, B, p) as the result of patching together the trivial fibrations $(U_i \times F_i, U_i, pr_1)$ by means of the transition functions ψ_{ji}. This construction is particularly appealing in the context of gauge theories, where one can think of global gauge transformations as the result of patching together local ones.

3.5. Principal fibrations (principal fiber bundles). The rather abstract definitions of the preceding sections become somewhat more palatable to physicists if we specialize the fiber to be a Lie group (more precisely, all fibers to be isomorphic to a Lie group). Then the group acts naturally on the space of the bundle (in distinction from the left action which is traditional in physics, it is convenient here to let the group act on the right, although everything works just as well with left actions; some texts are using the left action convention, e. g., Sulanke and Wintgen). We remind the reader that principal fibrations are motivated for us by gauge theory: a principal fibration is nothing else than a local gauge group on a manifold B. Recall the

3.5.1. Definition. A group G acts freely (without fixed points) from the right on a set E if for each $x \in E$ the subgroup which leaves the point invariant is the group identity e, i. e.,

$$S_x = \{g \in G: xg = x\} = \{e\}.$$

In other words, the mapping $x \mapsto xg$ is a bijection of G onto the orbit xG.

3.5.2. Definition. A principal fibration (principal fiber bundle) with base space B and structure group G is a quadruplet $\lambda = (P, G, B, \pi)$, where P is a manifold on which the Lie group G acts freely on the where singularities occur eliminated), the gauge group as the structure group. As we shall see, most gauge groups are trivializable principal fibrations.

iv) Let M be a manifold of dimension n, TM its tangent bundle. The basis in each tangent space can be changed by an action of the group GL(n, $\mathbb{R}$) of all nonsingular linear transformations. Letting the group element vary smoothly over a neighborhood of a point we obtain a local trivial principal fibration with GL(n) as the structure group, and by patching these together we obtain a GL(n)-principal fibration called the <u>bundle of frames</u> of the manifold M.

v) Consider a Möbius strip (E, S^1, p), with the fiber isomorphic to $\mathbb{R}$. A frame in this space is a non-zero real number, hence the bundle of frames of the Möbius strip is the Möbius strip with the zero-section removed, ant turns out to be a connected manifold, which points to the nonorientability of the Möbius strip (cf. Spivak,vol.II for more details).

<u>3.5.4. Morphisms.</u> Let $\lambda = (P, G, B, \pi)$ and $\lambda' = (P', G', B', \pi')$ be two principal fibrations. A <u>morphism</u> of λ into λ' is a triplet (f, φ, h), where $f:P \to P'$ and $h:B \to B'$ are morphisms of manifolds (C^∞-maps) and $\varphi: G \to G'$ is a Lie-group homomorphism, such that

$$\pi' \circ f = h \circ \pi \quad \text{and} \quad f(xg) = f(x)\varphi(g), \tag{3.6}$$

for $x \in P$, $g \in G$. The map h is determined completely by f, so that a morphism is determined by the pair (f, φ).

If $B = B'$, $h = Id_B$ we have a <u>B-morphism</u> compatible with φ, and if $G = G'$ and $\varphi = Id_G$ we say that we have a G-morphism compatible with h. A morphism which is both a B-morphism and a G-morphism is called a G-B-morphism. Any G-B-morphism is an isomorphism of manifolds $f:P \quad P'$ and the two fibrations λ and λ' are then said to be <u>isomorphic</u>.

<u>3.5.5. Trivializations.</u> The fibration $(B \times G, G, B, pr_1)$ is the trivial principal bundle over B with structure group G. Any isomorphism of a fibration $\lambda = (P, G, B, \pi)$ onto the trivial bundle over B

with structure group G is called a trivialization of λ .

A principal fibration is trivializable if and only if it admits a (global C^∞-)section $s:B \to P$.

Indeed, the mapping $f_s:\lambda \to (B \times G, G, B, pr_1)$ defined by

$$f_s^{-1}(b, g) = s(b)\cdot g \quad \text{for all } b \in B \text{ and } g \in G \qquad (3.7)$$

is a C^∞-bijection, hence an isomorphism. Thus there is a bijection between the set of trivializations of a principal fibration and the set of its global sections.

Since a gauge in a gauge-principal bundle is a global section of that fibration, to each gauge in a gauge theory there corresponds a different trivialization of the principal fibration. It is to be noted that the section has to be global, i. e., everywhere smooth. Gauges which are not everywhere smooth or everywhere defined occur in some physical applications.

3.5.6. Cocycle construction. The construction of fibrations by means of charts can be extended to principal fibrations. In this case the transition functions ψ_{ji} (we retain the notations of Section 3.4) are now group-valued mappings defined on $U_i \cap U_j$. More precisely, for an open covering (U_i) of B a cocycle (or C^∞-cocycle) on B with values in G is a family (ψ_{ij}) of mappings of the open set $U_i \cap U_j$ into G satisfying the cocycle identity (3.4). Two such cocycles ψ_{ij} and ψ'_{ij} are cohomologous if there exists a family (h_i) of C^∞-maps of U_i into G (for each index i), such that for all $x \in U_i \cap U_j$:

$$\psi'_{ij}(x) = h_i^{-1}(x)\psi_{ij}(x)h_j(x). \qquad (3.8)$$

In a principal fibration λ there always exist local sections over the open sets U_i, thus in $U_i \cap U_j$ there exists a unique mapping ψ_{ij} of $U_i \cap U_j$ into G such that it relates the local sections s_i and s_j:

$$s_j(b) = s_i(b)\cdot\psi_{ij}(b), \qquad b \in U_i \cap U_j. \qquad (3.9)$$

This family of mappings is a cocycle associated to the open covering of the base space of the fibration. Each of the local sections defines a local trivialization, and the trivializations f_i in different open sets U_i of the covering are related by the transition functions

$$f_i(x) = \psi_{ij}(\pi(x))f_j(x) \tag{3.10}$$

for $x \in \pi^{-1}(U_i \cap U_j)$.

Conversely, a cocycle associated to an open covering of B allows one to construct a principal fibration and a family of sections, as above, satisfying the conditions (3.9), (3.10). This principal fibration is unique up to isomorphism, and the cocycle condition can be considered as a patching condition of charts.

Two families of cohomologous cocycles determine G-B-isomorphic fibrations.

<u>3.6. Vector bundles associated to a principal fibration.</u> Instead of defining general fiber bundles associated to a principal fibration (i. e., bundles with fiber an arbitrary manifold), we restrict our atention to <u>associated vector bundles</u>, which are the objects for which the fields undergoing gauge transformations are sections. The general definition can be recovered by replacing in our definition vector spaces and representations by manifolds and left group actions (cf. Bourbaki, Sec. 6.5, or Dieudonné, (16.14.7)).

<u>3.6.1. Definition.</u> Let V be a vector space on which the group G acts on the <u>left</u> by a representation $r:G \to GL(V)$:

$$r: (g, v) \to r(g)v, \quad r(g_1g_2) = r(g_1)r(g_2), \quad g, g_1, g_2 \in G,$$
$$r(e) = I, \quad r(g^{-1}) = (r(g))^{-1}, \quad v \in V.$$

Then on the product of the principal bundle space P and V we define a <u>right action</u> of G as follows:

$$(x, v)\cdot g = (xg, r(g^{-1})v), \quad (x, v) \in P \times V. \tag{3.11}$$

The orbit space of this action, $(P \times V)/G$, i. e., the set of pairs (x, v) taken into each other by this action) will be denoted by $P \times^G V$ and is a manifold. For each orbit $z \in P \times^G V$ denote by $p(z) = p(x,v)$ the point of B equal to $\pi(x)$ for all $(x, v) \in P \times^G V$. Then the triplet $(P \times^G V, B, p)$ is a fibration called the vector bundle associated by r to the principal fibration $\lambda = (P, G, B, \pi)$, since it is locally trivializable and isomorphic to $U \times V$.

Any manifold E for which there is a morphism $\rho: P \times V \to E$ such that $\rho(xg, r(g^{-1})v) = \rho(x, v)$, and such that the quotient map $\bar{\rho}: P \times^G V \to E$ is a diffeomorphism is called a <u>vector bundle associated</u> to the principal fibration λ. The map ρ is called a <u>framing</u> of the associated vector bundle.

The quadruplet $(P \times V, G, E, \rho)$ is a principal fibration and the frame map satisfies the condition given above. One can define a projection $\pi_E: E \to B$ by

$$\pi_E(\rho(x, v)) = \pi(x), \quad x \in P, \; v \in V, \tag{3.12}$$

making the triplet (E, B, π_E) into a fibration. The mapping $\theta_x(v) = \rho(x, v)$ for fixed x is a vector space isomorphism (linear invertible map) of V onto the fiber E_b, and this map is equivariant under the action of G.

Finally, let s be a local section of P over the open set U in B. Then the frame map carries the trivialization of P over to E. Every section of the associated bundle $P \times^G V$ can be uniquely expressed in the form $\sigma: b \to \mathbf{r}(\mathbf{s}(b))\, v(b)$, where v(b) is a mapping from U into the typical fiber V. Since the zero representation always exists, the vector bundle will always admit the zero section as a global section, without necessarily being trivializable. On the other hand, the existence of a maximal set of linearly independent sections of a vector bundle (i. e., as many as the dimension of the fiber) allows one to deduce that there exists a global section of the associated principal fibration (or frame bundle) which is then trivializable, and so is the associated vector bundle. This remark is particularly relevant for gauge bundles.

<u>3.6.2. Example</u>. Consider the gauge principal fibration $(M \times SU(2), SU(2), M, pr_1)$, where M is Minkowski space. This is the gauge group of the Yang-Mills theory. Any principal fibration isomorphic to it will be trivialized by a global section, i. e., by a choice of gauge.

Every gauge leads to a different trivialization. Let V be the two-dimensional complex vector space of the isospinor representation. In the manner described above we can form the vector bundle described ab above, associated to the Yang-Mills gauge fibration. A section of this bundle is a classical isospinor field. Quantized fields will be defined later as operator-valued distributions on such sections.

One can consider higher-dimensional representations, by forming tensor products of the appropriate vector bundles , and decompose these into direct sums (Whitney sums), as one does with representations in general. We do not go into details here, but refer the reader to the literature.

<u>3.6.3. Vector bundles.</u> Vector bundles can be defined directly in terms of vector charts, i. e., triplets (U, ϕ, F) where U is an open set of the base space, F is a Banach space and ϕ is a bijection of $p^{-1}(U)$ onto $U \times F$ such that $p(\phi^{-1}(b, v)) = b$ for each $b \in B$ and $v \in F$. Just as for manifolds, one defines compatibility of vector charts, vector atlases and then a vector bundle as an equivalence class of vector atlases.

Sections of vector bundles (or local sections of vector bundles) have the additional property that they can be added and multiplied by scalars, thus forming vector spaces. Thus, the vector space of all smooth sections of a vector bundle E over the open set U of the base space will be denoted by $\Gamma(U, E)$. Each section of this vector space can be multiplied by any C^∞-function on U, and thus is a module over the ring of these functions.

We refer the reader to the literature for other properties of vector bundles and their sheaves of sections, since we will not make use of these properties in these notes.

4. CONNECTIONS, CURVATURE, AND HOLONOMY

4.0. Introduction. We have seen in Sections 2.2 and 2.3 that the electromagnetic vector potential and the Yang-Mills potential are to be interpreted as one-forms with values in the Lie algebra of the gauge group, and that the appropriate field strengths are two-forms with values in the Lie algebra, obtained by "gauge-covariant exterior differentiation" form the potentials. Such forms are known in differential geometry as connection one-forms and their curvature two-forms respectively, and form the subject matter of this chapter. There are several ways of introducing connections. We will take up the more abstract definition of a connection in a principal fibration in the following section, where the connection is introduced as a splitting of the tangent space of the principal fibration into a horizontal and vertical part, allowing the comparison of tangent vectors in different fibers. We will then discuss covariant differentiation and parallel translation in a general form, and end the chapter with a brief discussion of holonomy groups, i. e., with the transformations induced by parallel translation around closed loops in the base space. This is important in global aspects of gauge theories, which will be discussed in Chapter 6.

Before embarking on the more abstract definitions, we describe in this introduction the most elementary aspect of connections: their role in covariant differentiation. These definitions might suffice for some readers who wish to go on to the discussion of characteristic classes given in Chapter 5.

4.0.1. Connections as covariant differentials. Let (E, B, p) be a vector bundle, with fiber isomorphic to $\mathbb{R}^m$ over the n-dimensional manifold B (one may think of it as the vector bundle associated by an m-dimensional representation to a principal fibration, but the latter plays no role in this section). Denote by $\Gamma(B, E)$ the vector space of all C^∞-sections of E, i. e., of C^∞-maps $s: B \to E$ such that $p \circ s = \mathrm{Id}_B$.

A connection, or covariant differentiation, on the vector bundle E is a linear differential operator D from the space of sections of E to sections of the bundle $T^*(B) \otimes E$, where $T^*(B)$ is the cotangent bundle of B, i. e., the bundle formed out of the spaces of one-forms at all points of B:

$$D:\Gamma(B, E) \to \Gamma(B, T^*(B) \otimes E), \tag{4.1}$$

such that D satisfies the Leibniz rule: for any smooth section s(x) of E and any smooth function f we have:

$$D(fs) = df \otimes s + fDs. \tag{4.2}$$

Linearity means, as usual, that for two sections s_1 and s_2, and real costants a, b,

$$D(as_1 + bs_2) = aDs_1 + bDs_2. \tag{4.3}$$

The reader who is bothered by the tensor product sign may ignore it, remembering only that s takes values in E, which locally is like the product of a neighborhood U in B (which is locally $\mathbb{R}^n$) with $\mathbb{R}^m$, the fiber of E, whereas df is an n-dimensional one-form. Eq. (4.2) may be clearer if we introduce coordinates in B (summation over k):

$$\nabla_k(fs)dx^k = [\nabla_k f \cdot s + f\ \nabla_k s]dx^k. \tag{4.4}$$

Let $\{s_i\}$ $i = 1,\dots, m$ denote a local frame in U, i. e., a set of n linearly independent sections in a neighborhood U of B. The action of D on the sections s_i can be expanded in terms of the same local frame, with coefficients in $T^*(B)$, i. e.,

$$Ds_i = \sum \theta_i^j \otimes s_j. \tag{4.5}$$

The matrix of one-forms θ_i^j on U is called the connection one-form. It is convenient to follow Chern and rewrite equation (4.5) in matrix form. Let ${}^tS = (s_1,\dots,s_m)$ denote the row-vector of sections of the local frame, then the matrix Θ acts on the column-vector S according to (4.5):

$$DS = \Theta \otimes S. \tag{4.6}$$

If one thinks of the Yang-Mills theory of Sec. 2.3,

$$\Theta = \Theta_k dx^k, \quad \Theta_k = \sum A_k^\alpha \tau_\alpha$$

(cf. (2.76), showing that the connection one-form is actually a Lie-algebra valued one-form.

One can change the frame vector field by subjecting the sections s_i to local linear transformations:

$$S' = g(x)S, \qquad (4.7)$$

where $g(x) \in GL(m, \mathbb{R})$ is a matrix whose elements are C^∞-functions of $x \in B$. (If E is a vector bundle associated to a principal fibration by means of a representation r(G) of the structure group, g(x) is a representation of a local section of the principal fibration.)

Let Θ' denote the connection matrix relative to the new frame S', i. e., again in matrix notation,

$$DS' = \Theta' \otimes S'. \qquad (4.8)$$

Substituting (4.7) into (4.8) and using the properties (4.2) and (4.3) of the operation D (the elements of g(x) are functions, hence their differentials dg(x)):

$$D(g(x)S) = dg(x) \otimes S + g(x)DS =$$
$$= \Theta' g(x) \otimes S = (dg(x) + g(x)\Theta) \otimes S,$$

or, since the frame S is arbitrary and g(x) is invertible for each x:

$$\Theta' = dg \cdot g^{-1} + g\Theta g^{-1}, \qquad (4.9)$$

where for the sake of clarity we have left out the x-dependence of g. This is, up to notation, the transformation laws (2.78) of a Yang-Mills potential (with g^{-1} in place of g).

We now define the <u>curvature two-form matrix</u> by covariantly differentiating (4.5) or (4.6):

$$D(DS) = D(\Theta \otimes S) = d\Theta \otimes S + \Theta \otimes DS$$
$$= \{d\Theta - \Theta \wedge \Theta\} \otimes S,$$

where the minus sign in the last line is due to the fact that multiplication of one-forms is wedge-multiplication, and in order to maintain the rule for matrix multiplication, we must "interchange" the two connection matrices. We have thus defined the covariant exterior differential of the one-form matrix Θ which is the <u>curvature two-form</u> Ω:

$$\Omega = D\Theta = d\Theta - \Theta \wedge \Theta. \tag{4.10}$$

(In general, the covariant differential of a field of one-forms, i. e., a section of the cotangent bundle involves a minus sign, relative to the covariant differential of a vector field, as is well known for the case of a Riemannian connection.)

By differentiating both sides of Eq. (4.9) and defining Ω' in terms of Θ' by an equation like (4.10), it is easy to see that the curvature two-form has the simple transformation under a change of frame:

$$\Omega' = g\Omega g^{-1}, \tag{4.11}$$

i. e., transforms as a tensor. This is the familiar simple transformation property of the Yang-Mills field strength under a gauge transformation.

Taking the covariant differential of the two-form matrix Ω we obtain the Bianchi identity:

$$D\Omega = d\Omega - \Theta \wedge \Omega = 0, \tag{4.12}$$

which coincides with the Yang-Mills field equation.

It is sometimes convenient to replace the wedge product in (4.10) and (4.12) by a matrix commutator affected by a factor ½, in order to avoid double counting. In that case, the analogy with the Maurer-Cartan structure equation becomes particularly striking.

The reader should not be misled into thinking that the identity (4.12) means that the two-form matrix Ω is closed. In fact, the Bianchi identity is a differential equation for the connection one-form. However, in view of the transformation property (4.11) we shall see in Chapter 5 that the invariants of Ω can be obtained from the determinant

$$c(E, \Theta) = \det\{1 + i(2\pi)^{-1}\Omega\}, \tag{4.13}$$

which makes sense, since forms of even degree ahve commutative multiplication, and hence two-form matrices have determinants. The expression (4.12) is called the Chern form of the connection, it is closed, and its expansion into homogeneous forms leads to the Chern classes.

4.1. Connections in principal fibrations (principal connections). Motivated by the discussion in the preceding section we develop now a more general theory of connections in principal fibrations. This theory was developed from ideas of Elie Cartan by Kozsul and Ehresmann, and detailed discussions can be found in Kobayasi-Nomizu, Sternberg, Sulanke-Wintgen, Spivak, and Dieudonné, vol IV. Following Chern's lectures, we give several equivalent definitions of connections, all equivalent, in the hope that one or the other will be more accessible to the theoretical physicist.

4.1.1. First definition. Let $\lambda = (P, G, B, \pi)$ be a principal fibration over B with structure group G. A connection (or principal connection) on λ is a smooth family of vector subspaces H_p of $T_p(P)$ (the tangent space to the bundle space P at the point p), called horizontal subspaces, having the following properties:

(i) The subspace H_p is a direct summand in $T_p(P)$, i. e., each tangent vector to the bundle space has a unique decomposition into a horizontal part and a vertical part (tangent to the fiber at $\pi(p)$), or

$$T_p(P) = H_p \oplus V_p, \qquad (4.14)$$

where V_p is the tangent vector space to the fiber, i. e., the space of vertical tangent vectors).

(ii) The subspace H_p projects onto the tangent space to B at $\pi(p)$

$$\pi_* H_p = T_\pi(B). \qquad (4.15)$$

Here π_* is the derivative (1.1) of the projection map π.

(iii) The family of horizontal subspaces is invariant under the right action of the group G, i. e., for any $g \in G$

$$H_{p\cdot g} = H_p \cdot g = R_{g*} H_p, \qquad (4.16)$$

where R_{g*} denotes the right action of g on a tangent vector.

In other words, if we assume that the base space is n-dimensional and that the group G has q parameters, so that its Lie algebra, which is isomorphic to the vertical spaces V_p, has dimension q, then the tangent space $T_p(P)$ has dimension n + q. A connection is a smooth decomposition of T_p into a horizontal and vertical part, such that the horizontal subspace is right-invariant. The reader would do well to draw a picture of a trivializing local chart based on an open set U of B, and denote the point in P above U by (b, g). In this chart the tangent space $T_{(b,g)}(U \times G)$ is decomposed into $H_{(b, g)}$ and $V_{(b, g)}$ in such a way that $H_{(b, g)}$ projects onto $T_b(U)$ and $V_{(b, g)}$ is isomorphic, qua vector space, to the Lie algebra g of G. The mapping

$$c:T_b(U) \times P \to T_{(b, g)}(U \times G) \tag{4.17}$$

<u>lifts</u> any tangent vector to the base space into a horizontal tangent vector in $T_{(b, g)}$. In particular,

$$c((b, k), (b, g)) = ((b, k), P(b, g)\cdot k), \tag{4.18}$$

where $P(b, g)$ is a homomorphism of the vector space $T_b(U)$ into $T_g(G)$ and acts on the n-dimensional vector k in $T_b(U)$. The right invariance (4.15) implies

$$P(b, gg')\cdot k = (P(b, g)\cdot k)\cdot g' \tag{4.19}$$

In the special case when g = e, we will denote for later use

$$P(b, e) = Q(b) \in \mathrm{Hom}(T_b(U), g). \tag{4.20}$$

Any such homomorphism defines via (4.19) a homomorphism P and thereby locally, a connection.

The vertical vector space V_p is the kernel of the projection π_*. One can define a connection in a dual manner, by looking at those one-forms in $T_p^*(P)$ which annihilate horizontal vectors in H_p, forming the cotangent vertical forms V_p^*. This leads to the second defintion:

<u>4.1.2. Second definition .</u> Let $T_p^*(P)$ be the cotangent space of P at p, i. e., the vector space of all one-forms on $T_p(P)$. The space V_p^* is the space of all one-forms which vanish on horizontal vectors in H_p. We can call the smooth family V_p^* a <u>connection</u>. But the vector space V_p is isomorphic to the Lie algebra g, hence the one-forms in V_p^* can be considered as Lie-algebra valued one-forms (cf. Section 1.4), i. e., linear maps $\omega_p:T_p(P) \to g$, such that the subspace H_p is annihilated by them, i. e., $\langle \omega_p, X \rangle = 0$ for all vectors $X \in H_p$. The last relation

implies that

$$\langle \omega_{p\cdot g}, R_{g*}X \rangle = 0$$

for X in H_p, i. e., right invariance of the form. If X is a vertical vector, i. e., tangent to the fiber, the right action of the group element g on X is the adjoint action of the group element g on the vector X which may be considered an element of the Lie algebra of G:

$$\langle \omega_{p\cdot g}, R_{g*}X \rangle = \mathrm{Ad}(g^{-1})\langle \omega_p, X \rangle . \qquad (4.21)$$

On the fiber, i. e., in the vertical subspace, the one-form ω_p reduces to the Maurer-Cartan form $g^{-1}dg$, hence locally, in a neighborhood U in B, where the fibration is trivialized, ω_p must have the form

$$\omega_p = g_U^{-1}dg_U + \mathrm{Ad}(g_U^{-1})\,\theta_U(b, db), \qquad (4.22)$$

where $b = \pi(p)$ and θ_U is a $\mathfrak{g}$-valued one-form defined on U (related to the Q(b) defined in Eq. (4.20)). Under a right translation the form ω_p undergoes the transformation

$$\omega_{p\cdot g} = \mathrm{Ad}(g^{-1})\,\omega_p . \qquad (4.23)$$

The one-form ω (or the one-form θ_U) is called the connection one-form.

Conversely, if a Lie-algebra valued one-form is given which has the structure (4.22) and satisfies the condition (4.23), it defines a connection in the sense of definition 4.1.1 in the following manner. The horizontal space H_p is defined as the kernel of ω_p (i. e., the set of all tangent vectors in $T_p(P)$ for which $\langle \omega_p, X \rangle = 0$, and the vertical space contains all the vectors obtained by applying ω_p to an arbitrary vector in $T_p(p)$ [this is easily seen in terms of a chart, where a vector in $T_p(p)$ is n + q-dimensional and the connection form projects out the n-dimensional horizontal part, leaving a q-dimensional vertical vector]. Note that the form $g^{-1}dg$ is left-invariant and $\mathfrak{g}$-valued.

4.1.3. Third definition. This defintion is closest to the one familiar to physicists from general relativity, in terms of the transformation properties of the connection form under a change of chart. Thet U and V be two overlapping open sets in B, such that $b \in U \cap V$. The transition homomorphism corresponding to $U \cap V$ is ψ_{UV} (Sec.3.5.6) which is a cocycle with values in G. Then in the set $\pi^{-1}(U \cap V)$ the one-form θ_U defined in U is related to the one-form θ_V defined in V by means of the relation

$$\theta_U = \psi_{UV}^{-1}\, d\psi_{UV} + \mathrm{Ad}(\psi_{UV}^{-1})\,\theta_V. \tag{4.24}$$

Thus, a connection in the principal fibration is given by a Lie-algebra valued one-form θ_U defined in each open set of the covering {U, V,...} and satisfying the patching condition (4.24). We will call it a <u>local connection form</u>.

In the special case when $G = GL(q, \mathbb{R})$ ω and θ become matrices of one-forms, the adjoint representation becomes the representation by similarity transformations of matrices and (4.22) can be rewritten in the form

$$\omega = g_U^{-1}(dg_U + \theta_U g_U), \tag{4.25}$$

where g_U is a matrix in G (actually, a local section of the principal fibration). Let g_U, g_V be local sections, related by the transition homomorphism (3.9). Then the relation (4.24) becomes

$$(dg_U + \theta_U g_U) = \psi_{UV}^{-1}(dg_V + \theta_V g_V). \tag{4.26}$$

The expressions in parantheses have simple transformation properties under changes of charts, and lead us to the definition of covariant exterior differentials.

<u>4.2. Covariant differentiation. Curvature.</u> Generalizing the observation following (4.26) to an arbitrary principal fibration, we define the <u>exterior covariant differential</u> (the word exterior will be omitted in the sequel) as the horizontal projection of the exterior differential of a k-form on the bundle space P. Denoting by $h: T_p(P) \to H_p$ the projection on the horizontal subspace made possible at each point by the existence of a connection we define the covariant differential $D\xi$ of a k-form ξ by its values on a k + 1 - vector:

$$\langle D\xi, X_1 \wedge \dots \wedge X_{k+1} \rangle = \langle d\xi, hX_1 \wedge \dots \wedge hX_{k+1} \rangle, \tag{4.27}$$

where X_i are vectors in $T_p(P)$, i. e., vector fields on P. Since the right-hand side is invariant under the actions of the structure group, so is the left-hand side. If the k-form ξ is vertical (and if $k \geqq 2$), then $D\xi = 0$.

For the connection one-form the covariant exterior differential defines the <u>curvature two-form</u> Ω satisfying the <u>structure equation</u>:

$$\Omega = D\omega = d\omega + \tfrac{1}{2}[\omega, \omega], \tag{4.28}$$

where we have used the Lie-bracket notation in place of the wedge product, ant the factor $\tfrac{1}{2}$ takes care of the double antisymmetrization. In terms of the local form θ_U we have the local curvature form Θ_U:

$$\Theta_U = d\theta_U + \tfrac{1}{2}[\theta_U, \theta_U] \tag{4.29}$$

from where it can be shown that the local expression of is (cf.(4.22))

$$\Omega = \mathrm{Ad}(g_U^{-1})\Theta_U. \tag{4.30}$$

Under a change of chart the local expressions of the curvature two-form are related tensorially:

$$\Theta_U = \mathrm{Ad}(\psi_{UV}^{-1})\Theta_V, \tag{4.31}$$

showing that all invariants of the connection will be obtained via the curvature.

In order to prove the structure equation (4.28) one proceeds as follows. Apply the left-hand side to a bivector $X_p \;\; Y_p$, and consider the three possible cases: i) $X_p, Y_p \in H_p$, ii) $X_p \in H_p, Y_p \in V_p$, iii) $X_p, Y_p \in V_p$. In case i), since the form ω is vertical (4.28) reduces to the definition (4.27). In case ii) extend X_p to a vector field Z on B (i. e., $\pi_*(X_p) = Z$, and denote by A the element of g corresponding to the vertical vector field Y_p. Then

$$\langle d\omega, A \wedge Z\rangle = A\langle \omega, Z\rangle - Z\langle \omega, A\rangle - \langle \omega, [A, Z]\rangle$$
$$= -[\langle \omega, A\rangle, \langle \omega, Z\rangle] + \langle D\omega, A \wedge Z\rangle.$$

The first term in the right-hand side of the first row vanishes since Z is horizontal, and the second, since $\langle \omega, A\rangle$ is constant. The last term is equal to the right-hand side of (4.27), and comparison with the last row proves the structure equation in this case. Finally, in case iii) the calculation is almost identical, except that both vectors are now vertical.

When the base space reduces to a single point, so that the bundle space may be identified with the Lie group G, there is only one connection, and the connection form reduces to tha Maurer-Cartan form, satisfying the structure equation

$$d\omega + \tfrac{1}{2}[\omega, \omega] = 0, \tag{4.32}$$

which shows that the curvature form vanishes (a connection with vanishing curvature form is called flat).

4.2.1. Bianchi identity. The curvature form Ω satisfies the Bianchi identity $D\Omega = 0$. Indeed, the structure equation and the definition of covariant differentiation yield:

$$D\Omega = D(d\omega) + \tfrac{1}{2}D([\omega , \omega]) = 0, \tag{4.33}$$

the first term in the right-hand side being zero because $dd\omega = 0$, and the second being zero since $[\omega, \omega]$ is a vertical form.

In the case of the structure group $GL(q, \mathbb{R})$, taking the first row of the matrix equation (4.26) and denoting the row vectors of g_U, g_V respectively by s_U, s_V, we obtain

$$Ds_U \psi_{UV} = Ds_V , \tag{4.34}$$

where

$$Ds_U = ds_U + \theta_U s_U \tag{4.35}$$

We can make the first row vectors of the matrices g_U into a vector bundle associated to the GL-principal fibration; then (4.33) defines the covariant differentiation of sections of that bundle,, as described in Section 4.0.1. (one should not confuse the matrix Θ used there with the local curvature two-form matrix (4.29)!). In the next section we discuss connections in associated vector bundles in more generality.

4.3. Connections in associated vector bundles (Linear connections). Recall the defintion of an associated vector bundle, Sec. 3.6. Suppose we have a principal connection in the sense of definition 4.1.1. in the principal fibration P to which the vector bundle $E = P \times^G V$ is associated, i. e., a smooth splitting of the tangent spaces $T_p(P)$ into vertical and horizontal parts V_p, H_p, satisfying (4.14) - (4.16).

Let $b \in B$ and $u_b \in E_b$ be a vector in the fiber over b of E. There exists an element $p_b \in P_b$ of the principal fibration (locally represented by the pair $(b, g) \in B \times G$) and a vector $v \in V$, such that $u_b = \theta_b(v)$ where θ_b is the linear isomorphism of vector spaces defined after Eq. (3.12) (not to be confused with a connection form).

Then the family of horizontal subspaces $H = H_p$ is mapped into a family of subspaces in T(E) by the map σ_p induced by the above construction between P and E, yielding a horizontal space $D_v = \sigma_p^*(H_{p_b})$. It can be shown that this map is independent of the choice of p in the fiber P_b, since replacing p_b by $p_b \cdot g$ and v by $r(g^{-1})v$ leads to the same result, hence this mapping is invariant under the right action of G. The set of all spaces D_v determines a differential system on E and the space D_v is the complement to the tangent space to the fiber. Any curve in the base space, C, lifts to a horizontal curve γ in P (see the following section for details), and $\sigma_p\gamma$ is a solution of the differential system.

More specifically, in a chart containing the point b and a tangent vector $k_b \in T_b(B)$ we define a linear connection C in E by

$$C_b(k_b, u_b) = P_b(k_b, p_b) \cdot v, \tag{4.36}$$

where P_b is a homomorphism of the vector space $T_b(B)$ into $T_g(G)$ defined in Eqs. (4.18) - (4.20). $C_b(k_b, u_b)$ is independent of the choice of the pair (p_b, v) such that it yields the same u_b. If π_E denotes the projection of E on B we have for the tangent maps

$$T(\pi_E)C_b(k_b, u_b) = T(\pi)\ _b(k_b, p_b) = k_b, \tag{4.37}$$

i. e., both horizontal subspaces, in the vector bundle and in the principal bundle, project onto the tangent space of the base space. The mapping $u_b \in C_b(k_b, u_b)$ is linear from E_b to $(T(E))_{k_b}$, since all the ingredients are linear. In the chart we have then, with the notation (4.20):

$$C_b((b, k), (b, v)) = ((b, v), -\Gamma_b(k, v)) \tag{4.38}$$

$$\Gamma_b(k, v) = -\rho_*(Q(b) \cdot k) \cdot v;$$

Here ρ_* is the tangent map to ρ defined in section 3.6, and is a homomorphism of g into $g\ell(V)$, the endomorphisms of the vector space V.

To connect this construction of a connection in a vector bundle, which is somewhat abstract, with the defintion in terms of covariant differentiation given in the introduction, we first note that locally any section of the principal fibration $s: U \in P_U$ determines a section in the associated vector bundle E via the relation

$$S(b) = r(s(b))\cdot\zeta(s(b)), \tag{4.39}$$

where r is the representation on which E is based and ζ is a C^∞-map of a neighborhood of $p_b = s(b)$ into $E_b = V$. Then the covariant derivative of the section S in the direction of the tangent vector k_b $T_b(B)$ is obtained as follows. The tangent vector $T_b(S)\cdot k_b$ in E has both horizontal and vertical components. We use the connection C to extract the horizontal component, and thus to compare vectors tangent to the fibers at neighboring points of B (separated by the displacement k_b). The vector

$$T_b(S)\cdot k_b - C_{\pi_E(S(b))}(T_b(\pi_E\circ S)k_b,\ S(b))$$

is in the tangent space $T_{S(b)}(E)$, which is isomorphic to the Lie algebra $g = g\ell(V)$. We now calculate $T_b(S)$ by taking the tangent linear maps of Eq. (4.39):

$$\begin{aligned} T_b(S)\cdot k_b &= T_b(r(s(b))\cdot k_b\cdot\zeta(s(b)) \\ &\quad + r(s(b))\cdot T_{s(b)}(\zeta)\cdot(T_b(s)\cdot k_b) \end{aligned}$$

Denoting by h the vector which projects into k_b: $T(\pi)\cdot h = k_b$, $h = T_b(s)\cdot k_b \in T_p(P)$, then $T_{s(b)}(\zeta)\cdot h$ is a vector in $T_{\zeta(p_b)}(V)$. Finally, denoting by $\tau_{S(b)}$ the canonical linear bijection of $T_{S(b)}(E_b)$ onto V, we obtain the formula for covariant differentiation of a section, as the vector in V

$$\nabla_{k_b}\cdot S = \tau_{S(b)}(T_b(S)\cdot k_b - C_b(k_b,\ S(b)), \tag{4.39}$$

which in view of the preceding discussion can be rewritten in the form

$$\nabla_{k_b}\cdot S = \bar{\rho}(\theta_h\cdot\zeta + \tau_{\zeta(p_b)}(\omega_b\cdot h)\cdot\zeta(p_b)). \tag{4.40}$$

Here ω_b is the connection one-form defined in (4.1.2), and $\bar{\rho}\rho$is the frame map of E defined in 3.6.1. It is easy to check that ∇ is a covariant derivative, satisfying the requirements (4.2),(4.3). In Eq. (4.40) $\theta_h \cdot \zeta$ denotes the Lie derivative of ζ equal to $\tau^{-1}(T_{s(b)} \cdot h)$. (For the definition of the Lie derivative, which will not be used in any other place, cf. Dieudonné, Section 17.14.7, or any other text). For details of the calculation presented here, cf. Dieudonné, Sec. 20. 5). For most of the remainder of these notes, it suffices to think of a connection in terms of the definition given in the introduction, or in terms of the connection one-form in the principal fibration.

The covariant differential can be expressed in terms of the covariant derivative of the action of the 1-form ϕ on vectors:

$$D\phi(X \wedge Y) = \nabla_X\langle \phi, Y\rangle - \nabla_Y\langle \phi, X\rangle - \langle \phi, [X, Y]\rangle \tag{4.41}$$

and is easily generalized to higher forms.

The curvature of the linear connection can be defined as before, as the exterior covariant differential of the connection, or via the commutation properties of the covariant derivative. We refer the reader to the literature for details.

<u>4.4. Parallel translation.</u> A <u>path</u> in the manifold M is a continuous mapping of the interval [0, 1] into B which is piecewise smooth, $t \to \gamma(t)$. A <u>loop</u> is a path with $\gamma(0) = \gamma(1)$.

Let ω be a connection in the principal fibration $\lambda = (P, G, B, \pi)$. A path $\gamma(t)$ in P is <u>horizontal</u> if its tangent vectors $d\gamma/dt$, including the one-sided tangent vectors at the endpoints of the path are horizontal. A path $\gamma(t)$ in P is a <u>lifting</u> of the path $\sigma(t)$ in B if $\pi(\gamma(t)) = \sigma(t)$. Since the group action in P is vertical, the right translate of a horizontal path $\gamma(t)$ is also horizontal (we denote this new path by $R_g\gamma(t)$. Not every mapping of a manifold X into B can be lifted to the bundle P, but paths are an exception. In fact one can easily prove that to any path $\sigma(t)$ in the base space and any point p_0 in P <u>there exists a unique horizontal lifting</u> $\gamma(t)$ such that if

$p_0 = \gamma(t_0)$, then $\pi(p_0) = \sigma(t_0)$. In other words, through each point p_0 of the fiber $\pi^{-1}(\sigma(t_0))$ there passes one and only one horizontal lift of the path $\sigma(t)$. For the proof of this fact, we refer the reader to the literature, e. g., Kobayasi and Nomizu.

<u>4.4.1. Definition.</u> Let E be a vector bundle associated to the principal fibration λ, and let (t) be a path in B with a horizontal lift p = (t), such that $p_0 =$ (0), $p_1 =$ (1). Then the connection in the principal fibration induces an isomorphism depending on the path of the fibers over $b_0 =$ (0) and over $b_1 =$ (1) in E, i. e., a linear mapping of the the two vector spaces into each other which is invertible and called <u>parallel translation</u> of vectors in E.

Another way of defining parallel translation is by requiring that the covariant derivative of vectors parallel-transported along a path vanishes.

If the vector bundle is the tangent bundle of a manifold, the connection is called a <u>linear connection</u>, and the integral curves for which the tangent vectors are parallel dispalced with respect to one another are called <u>geodesics</u>. We leave it to the reader to reestablish the relation with these concepts familiar from general relativity.

Finally, <u>parallel translation of fibers in a principal bundle</u> can be defined as the group isomorphism between the fibers induced by the connection along a horizontal lift of a path in the base space. In this case there is a path g(t) in the structure group G such that

$$\langle \omega(\gamma(t)),\gamma'(t)\rangle = -g(t)g'(t)\,, \tag{4.42}$$

where ω is the connection one-form at a point of the horizontal lift, and the prime denotes differentiation with respect to t. (Cf. Dieudonné, vol. IV, Problem 20.2 # 3.)

It is useful to introduce a formalism similar to the Dyson time-ordering in S-matrix theory, to represent the linear mapping which effects the parallel translation along a path. Thus, assuming that the vector $u(t_0)$ is parallel translated along the curve $\gamma(t)$ into the vector $u(t_1)$, such that, if we denote a basis vector in the tangent space to the base space by e_i and the covariant derivative of that vector u in the direction e_i by ∇_i, we have the condition of parallel translation

$$\nabla_i u = \partial_i u + \omega_i u = 0, \tag{4.43}$$

where ω_i is the (antihermitian) matrix corresponding to the connection one-form in the representation r(G) and ∂_i is the ordinary partial derivative, corresponding to the change in the horizontal part of the vector field between neighboring points in the limit of zero displacement. If we denote by iT^a the (antihermitean) generators of the Lie algebra in the representation r(G),then (4.43) can be rewitten in a form analogous to Eq. (2.77)

$$\partial_k u + i \sum A_k^a T^a u = 0, \tag{4.44}$$

where A_k^a are the coefficients of the connection form ("gauge fields"). The differntial equation (4.44) can be integrated along the path $\sigma(t)$ in B, if we introduce the "Dyson path-ordering" operator P, analogous to the time-ordering operator in S-matrix theory (since the matrices do not commute , or in the quantum field theory case, the field operators do not commute, this ordering is necessary; in the mathematical literature the right-hand side of Eq. (4.45) is known as a "product integral", cf. e. g., Gantmakher, Theory of Matrices): (sum over a,k)

$$u(b_1) = P \exp(-i\int A_k^a(b) T^a db^k) u(b_0). \tag{4.45}$$

The ordered exponential can also be understood as a limit of products of factors of the form $(1 - iA_k^a(b_j)T^a \Delta b_j)$ as $\Delta b_j \to 0$, b_j being points in B along the path.

If the path is a closed loop, the parallel transport will subject the vector $u(b_0)$ to a linear transformation

$$u'(b_0) = P \exp(-\oint \omega) u(b_0), \tag{4.46}$$

where we have used the connection form, for brevity. If there were no path-ordering, we could reduce the integral by Stokes' theorem to an integral of the curvature two-form Ω; cf. below, Section 4.5.1.

4.5. Holonomy groups and the Ambrose-Singer theorem. In this section we review briefly the interesting concept of holonomy and refer the reader for details to the literature (e. g., Lichnerowicz, Kobayasi and Nomizu, or for the classical theory, Vranceanu, vol. III).

We again consider a principal fibration $\lambda = (P, G, B, \pi)$ and a principal connection ω defining parallel translation of fibers, which are here group homomorphisms. Let $\gamma(t)$ be a closed loop in the base space, i. e., a path for which the initial and final points coincide. According to the prceding section, to each such path starting from a fixed point b_0 in B there corresponds an isomorphism o the fiber P_{b_0} (which is isomorphic to G) onto itself. We denote this isomorphism associated to the loop γ by $[\gamma]:P_{b_0} \to P_{b_0}$. Since loops starting at a given point form a group (under consecutive traversal as product, traversal in reverse direction as inverse; when forming the product it is convenient to double the parameter, so that the final parameter runs again over the interval [0, 1]), we obtain a group of isomorphisms of the fiber P_{b_0} onto itself (the reader will verify easily that to composition of the loops corresponds the product of isomorphisms and to the inverse loop the inverse homomorphism. The group of all the fiber isomorphism at a given point is called the holonomy group K_{b_0} at b_0. If one restricts one's attention to loops which can be contracted to a point -- so called null-homotopic loops -- one obtains a subset of K_{b_0} called the restricted holonomy group at b_0 and denoted by $K^0_{b_0}$. The same concept can be introduced in an associated vector bundle, except that here the holonomy group will be a group of linear isomorphisms of the fiber onto itself, and in fact will be given by the representation $r(K_{b_0})$ induced by the representation $r(G)$.

Taking another base point b_1 in B which can be connected to b_0 by a path, we can construct the holonomy group K_{b_1} at b_1 and relate the latter to K_{b_0} as follows. Consider the loop at b_0 consisting of the path c from b_0 to b_1, the loop γ at b_1 and the path $-c$ from b_1 to b_0

The appropriate fiber isomorphism of fibers is then $[c][\gamma][c]^{-1}$, where [c] denotes the isomorphism of fibers given by parallel translation along c. Hence the groups K_{b_0} and K_{b_1} are isomorphic and we can talk about <u>the holonomy group</u> of the connection, if the base space B is connected. If the base space is simply connected, i. e., every loop in it is contractible, the holonomy group and the restricted holonomy group coincide.

We now show that the holonomy group can be considered a Lie subgroup of the structure group G. Indeed, since the fiber is isomorphic to G, we can associate by $[\gamma].p = p.g_\gamma$, $p \in \pi^{-1}(b_0)$, an element g of G to each element $[\gamma]$ of K_{b_0}. The image of this injective homomorphism in G will be denoted by the same letter K_p. It is easy to see that

$$K_{p\cdot g} = g^{-1}K_p g. \tag{4.39}$$

Similar statements hold for the restricted holonomy group. The holonomy group K_p is a Lie subgroup of G (if B is paracompact, i. e., admits locally finite open coverings, which we have tacitly assumed throughout), and K_p^0 is a normal soubgroup. Moreover, the quotient K_p/K_p^0 is countable (it has obviously some relation to the connectivity of B). The proof is not simple, particularly the fact that K_p^0 is a Lie group, and we refer the reader again to Kobayasi and Nomizu for details.

The set of all points in P which can be connected to a point p by a horizontal path is the reduction of the principal fibration to the holonomy group K_p, and the holonomy group of the reduced connection is again K_p, i. e., in such a reduced bundle the structure group and the holonomy group of the connection coincide. Such a reduced fibration is sometimes called the <u>holonomy bundle</u> of the connection ω at p.

The Lie algebra k_p of the holonomy group is called the <u>holonomy</u> algebra of the connection. If B is simply connected the restricted holonomy group which coincides with the holonomy group is the unique Lie subgroup of G determined by the Lie algebra k.

4.5.1. The Ambrose-Singer holonomy theorem[2] relates the holonomy algebra of a connection to the curvature form of the connection. The curvature two-form Ω of a connection is obviously related to parallel translation around the boundary of a two-cell, i. e., around an infinitesimal parallelogram in the base-space. But parallel translation around an infinitesimal loop must give rise to elements of the holonomy group in the neighborhood of the identity of the group, i. e., to an element of the holonomy algebra. This is made more precise in the following form of the Ambrose-Singer theorem:

Let $\lambda = (P, G, B, \pi)$ be a principal fibration over a paracompact connected base-manifold B, with a connection form ω and curvature two-form Ω. Then the holonomy algebra k_p at a point $p \in P$ is spanned by all the elements $\langle \Omega, X \wedge Y \rangle_q$, where q is a point obtained from p by parallel translation along any path in B (i. e., q is a point in the holonomy bundle of p), and X, Y are horizontal tangent vectors in H_q.

The proof is simple if one assumes the bundle reduced to the holonomy bundle. In that case, the vector space spanned by the elements described above is an ideal of the Lie algebra k_p. By splitting the elements of this ideal into vertical and horizontal components, and taking into account the properties of integral curves of these vector fields, one then proves easily that the dimension of the ideal is equal to the dimension of the holonomy algebra, and therefore coincides with it. For details of the proof the reader is referred to Kobayasi and Nomizu, or to Sulanke and Wintgen.

If the curvature of a principal fibration vanishes the bundle is called flat (parallelizable), and the connection is trivial (in gauge theory language: a pure gauge field, which can be gauged away, i. e., the connection form is a pure Maurer-Cartan form associated to a section of the bundle).

If the holonomy group of a connection is abelian the connection is such that its curvature form spans an abelian Lie algebra.

5. AN INTRODUCTION TO CHARACTERISTIC CLASSES

5.0. Introduction. The theory of characteristic cohomology classes is one of the most beautiful achievements of mathematics of the past forty years. Unfortunately, the presentation in most mathematics texts is difficult to read for physicists, since the theory is usually presented in an axiomatic form with little or no motivation. I have found the lectures by Chern and the article by Bott and Chern [8] particularly clear and easy to follow. More information can be found in Hirzebruch, Husemoller and Milnor and Stasheff. Characteristic classes of vector bundles have acquired a particular significance for gauge theories in connection with the discoveries of "pseudoparticle solutions" or "instantons" by Polyakov and coworkers and 't Hooft, but as is usually the case when a new concept is introduced into physics, there is still a lot of confusion as to the correct interpretation. It is the purpose of this short chapter to motivate the concepts and introduce the basic definitions of Chern classes of complex vector bundles. The Pontryagin classes which are characteristic for real vector bundles (e. g., for gravitation, or for the case of orthogonal gauge groups) appear as special cases. We also briefly dicuss various "integrality" theorems, which associate integers to the integrals of certain forms, integers which have played a role in the recent discussion of "topological charges", as well as possibly in quantization problems.

We have seen in Section 1.3 how cohomology groups are introduced in the discussion of differential forms, and have sketched some applications to electromagnetism in Section 2.4. In those sections there appeared certain integers (the Euler-Poincaré characteristic, or the charge in units of an elementary charge). In this chapter, and in the applications of the following chapter, we shall relate these integers to integrals of certain forms - mainly curvature forms of connections or products of such forms with themselves, since we are integrating on four-dimensional manifolds - obtaining various generalizations of the classical Gauss-Bonnet formula (the Gauss-Bonnet-Chern-Allendoerfer-Weil formula) which relates the Euler characteristic of a surface to integrals of the various curvatures of that surface.

The theory of characteristic classes is associated to the names of Whitney, Stiefel, Pontryagin and Chern, and was developed further by Weil, Bott, Thom and many others. Whitney and Stiefel introduced characteristic classes in 1935. Stiefel studied the homology classes determined by the tangent bundle of a smooth manifold and invented cohomology theory, whereas Whitney discussed the case of sphere bundles, which have the advantage of having compact fibers. Pontryagin constructed the classes which bear his name by studying the homology of so-called Grassmann manifolds (the manifolds of all q-dimensional linear spaces through the origin of a q + n-dimensional Euclidean space), i. e., characteristic classes of vector bundles associated to orthogonal structure groups (Pontryagin's work goes back to 1942). In 1946 Chern defined characteristic classes for complex vector bundles, and showed that complex Grassman manifolds are easier to understand than the real ones. Of the integrality theorems mentioned earlier, in addition to the Gauss-Bonnet et al. theorem, Hopf had discovered in 1927 that the number of zeroes of a smooth vector field on a compact oriented manifold is equal to its Euler characteristic; Thom and Wu proved that the integrals of the highest-dimensional Chern class equals the Fuler characteristic, and Hirzebruch constructed a number associated to the tangent bundle of a 4k-dimensional real manifold (compact and oriented) called the L-genus, and proved that it is equal to another integer, called the signature. In the case of 4-dimensional manifolds it turns out to be equal to one-third of the integral of the first Pontryagin class of the manifold. There have been many generalizations, such as the extension of Hopf's result to sections of complex vector bundles by Bott and Chern[8] and the various "index theorems", the most famous of which is the Atiyah-Singer index theorem, which relates the index a manifold to the index of an elliptic differential operator (the Laplacian) on that manifold. Unfortunately, we do not have the possibility of discussing these developments, which may be very important for physics, in more detail in these notes, but hope to return to them elsewhere.

Our discussion of characteristic classes is rather heuristic and follows mainly the ideas of Weil as presented by Bott and Chern. The reader interested in more detail is referred to the book by Milnor and Stasheff (Appendix) or any of the other books listed. In view of their importance for gauge theories, where the vector bundles are complex and the structure groups are unitary, we restrict our attention to Chern classes, and only briefly mention how they are related to Pontryagin classes. This construction is due to Chern, Weil and Bott, and is taken essentially from the paper by Bott and Chern [8].

5.1. Curvature and Chern classes. We return to the setting of Section 4.0, considering a complex vector bundle E over a (compact) manifold B. We denote by A(B; E) the graded ring of all E-valued differential forms on B, and by d the differential operator which produces the grading (i. e., the operator which takes us from the q-forms in A^q to the $q + 1$ - forms in A^{q+1}, such that $d^2 = 0$. If $\Gamma(B, E)$ denotes the smooth sections of E then A(B; E) can be represented as the tensor product of A(B) (the complex-valued differential forms) with $\Gamma(B, E)$.

We recall that a connection in the vector bundle E is a linear operator D, (4.1), satisfying the Leibniz rule (4.2), and if s_j denotes a frame of E, i. e., a set of linearly independent sections, then D can be characterized by the connection matrix $\theta_i^j \in A^1(U)$, where U is an open neighborhood in B. The curvature matrix Ω defined by (4.10) is a matrix of two-forms on U relative to the given frame. Since even forms have commutative wedge-products (1.10), it makes sense to define determinants of matrices formed out of even forms.

In particular, we can form the determinant of the matrix $1 + i\Omega/2\pi$ which leads to all the invariants (under the action of the structure group via the representation $r(G)$) of the curvature matrix:

$$c(E, D) = \det\{1 + i\,\Omega/2\pi\}. \tag{5.1}$$

This is an element of $A(U)$, the set of all (not necessarily homogeneous) differential forms on U , which apparently depends on the choice of frame in U. The factor $1/2\pi$ will lead to integer values of the integrals of the Chern forms, and is tantamount to a choice of "solid angle". We first show that the Chern form (5.1) does not in fact depend on the frame, and hence defines a global form on B, rather than a local one in U only (once we know that $c(E, D)$ is independent of the frame, we cover B with open sets, and use different trivializations in each; on the overlaps of the open sets, the tensorial character of the curvature (4.11), together with the invariance of a determinant under a conjugation, imply the globality of $c(D, E)$). If we subject the local frame S to a linear transformation $S' = AS$ (we use the matrix notation (4.7)), the connection matrices are related by Eq. (4.9):

$$dA + A\theta(S, D) = \theta(S', D), \tag{5.2}$$

and the curvatures are intertwined by A:

$$A\Omega(S, D) = \Omega(S', D)A, \tag{5.3}$$

where we have made explicit the dependence on the frame S, and the connection D. Recalling the invariance of a determinant under a conjugation, we obtain the independence of the form $c(E, D)$ of the choice of S and hence on the choice of the covering U.

Next we will show that $c(E, D)$ is a closed form $dc(E, D) = 0$, and hence defines a cohomology class, and then we show that the cohomology class does not depend on the connection D, but only on the bundle E itself, and is thus a characteristic class. The transformation law (5.3) is characteristic for all matrix-valued differential forms on B, i. e., the curvature matrix $\Omega(S, D)$ is an element of $A^2(B;\mathrm{Hom}(E,E))$ which will be denoted, following Bott and Chern, by $K[E, D]$.

A k-linear function over M_n, the space of n by n complex matrices, $F(X_1,\ldots,X_k)$ is called invariant if for any $A \in r(G)$

$$F(Ad(A)X_1,\ldots, Ad(A)X_k) = F(AX_1A^{-1}, \ldots , AX_kA^{-1}) = F(X_1,\ldots,X_k). \quad (5.4)$$

This function can be extended to matrix-valued forms on $U \times B$ by setting

$$F_U(X_1\omega_1,\ldots,X_k \omega_k) = F(X_1,\ldots,X_k)\omega_1 \wedge \omega_2 \wedge\ldots\wedge\omega_k,$$

where $\omega_i \in A(U)$ are differential forms on U with values in C.

Now consider k matrix-valued differential forms $\xi_i \in A(B; \mathrm{Hom}(E,E))$ and let F be as above. Then , given a frame S over U, we obtain an invariant function $F_U(\xi_1,\ldots,\xi_k)$ which is the complete polarization of the "polynomial" $F_U(\xi ,\ldots,\xi) = F_U((\xi))$ (the notation of Bott and Chern). In particualr, for a connection D, we have the well defined forms $F((K[E, D]))$ and $F((1 + iK[E, D]/2\pi))$ in A(B) and the Chern form is a special case of the latter, since the determinant is an invariant polynomial:

$$c(E, D) = \det\{1 + iK[E, D]/2\pi\}. \quad (5.5)$$

Now we use the identity (5.4) in its infinitesimal form and the Bianchi identity (4.12) to show that for any invariant k-linear function F the form $F((K[E, D]))$ is closed, and hence belongs to a cohomology class, and then that the cohomology class is independent of the choice of the connection D. For this we rewrite (5.4) in its infinitesimal form (obtained by differentiation with respect to A):

$$\sum_{j=1}^{k} F(X_1, \ldots, [X_j, A], \ldots,X_k) = 0, \quad (5.6)$$

where the bracket is the commutator of the matrices. This identity can be extended to F_U for p-form matrices, if one defines the bracket of a p-form with one of a q-form (cf. Section 1.4)

$$[X^p, Y^q] = X \wedge Y - (-1)^{pq}Y \wedge X. \quad (5.7)$$

Thus, if Y is a q-form and $X_1,\ldots,X_k$ are $p_1,\ldots,p_k$-forms with values in M_n (or in a Lie algebra, for that matter), (5.6) becomes:

$$\sum_{j=1}^{k} (-1)^{q(p_1 + \ldots + p_{j-1})}F(X_1,\ldots,[X_j, Y],\ldots,X_k) = 0. \quad (5.8)$$

Since the bracket is a derivation, just like the exterior differential, it follows that the exterior differential of F_U is

$$dF_U(X_1,\ldots,X_k) = \sum_{j=1}^{k} (-1)^{p_j + 1 + \cdots + p_k} F_U(X_1,\ldots,dX_j,\ldots,X_k), \tag{5.9}$$

where the sign is determined by the antisymmetry of the exterior differential.

If now F is any invariant k-linear matrix form and D a connection with curvature form K[E, D], then the Bianchi identity implies that F((K[E, D])) is <u>closed</u>. This can be proved locally, i. e., in a neighborhood U, with a particular frame S. Since Ω is even, the sign factor in (5.9) disappears and we have:

$$\begin{aligned} dF_U((\Omega(s, D))) &= \sum_{j=1}^{k} F_U(\Omega,\ldots,d\Omega,\ldots,\Omega) \\ &= \sum_{j=1}^{k} F_U(\Omega,\ldots,[\theta, \Omega],\ldots,\Omega) \\ &= 0 , \end{aligned} \tag{5.10}$$

where we have used the Bianchi identity $d\Omega = [\theta , \Omega]$ and the identity (5.8) with $q = 1$, $p_i = 2$. Note that $F_U((\Omega))$ is a form of degree 2k. Since it is a closed form, it defines a cohomology class, i. e., a subspace of 2k-forms differing from each other by an exact form, i.e., an element of $H^{2k}(B, \mathbb{R})$ (cf. Section 1.3). To prove that this class does not depend on the connection, and depends only on the function F one makes use of the folowing "homotopy" argument.

We define the one-parameter family of connection forms "interpolating" between the two connection forms θ_0 and θ_1 by:

$$\theta_t = \theta_0 + t\alpha , \quad \alpha = \theta_1 - \theta_0, \tag{5.11}$$

and the appropriate curvature forms

$$\begin{aligned} \Omega_t &= d\theta_t - \tfrac{1}{2}[\theta_t, \theta_t] \\ &= \Omega_0 + t(d\alpha - [\theta_0, \alpha]) - \tfrac{1}{2}t^2[\alpha , \alpha]. \end{aligned} \tag{5.12}$$

Differentiating with respect to t we obtain after inserting in $F((\Omega_t))$

$$\begin{aligned} (1/k)(d/dt)F_U((\Omega_t)) &= F_U(d\alpha - [\theta_t,\Omega],\Omega_t,\ldots,\Omega_t) \\ &= dF_U(\alpha, \Omega_t,\ldots, \Omega_t), \end{aligned} \tag{5.13}$$

where we have made use of (5.8). Integrating with respect to t from 0 to 1, we obtain

$$F((K[E, D_1])) - F((K[E, D_2])) = kd\int_0^1 F(\alpha, K[E, D_t], \ldots, K[E, D_t])\, dt, \tag{5.14}$$

where D_t denotes the connection with the connection form θ_t, and similarly for D_0 and D_1. This means that the two 2k-forms in the left-hand side differ by an exact form, i. e., are in the same cohomology class. Moreover, setting $D_0 = 0$ in (5.14) (such a trivial connection always exists), one can express $F((K[E, D]))$ as a coboundary.

If we denote by $w(F)$ the cohomology class represented by the closed form $F((K[E, D]))$, we establish a homomorphism between the ring of invariant k-linear functions and the cohomology ring of the bundle, homomorphism which is known in the literature as the <u>Weil homomorphism</u>.

To obtain the Chern classes, we specialize the function F to be the determinant (5.5). Then expanding the Chern form $c(E, D)$ into homogeneous forms (all of even degree)

$$c(E, D) = 1 + c_1(E) + \ldots + c_k(E) + \ldots \tag{5.15}$$

(the sum is finite, since the base space of the bundle is finite-dimensional), we obtain the <u>Chern classes</u> $c_k(E)$ of the vector bundle E. Our calculation yielded cohomology classes with complex coefficients, but it can be shown that with the chosen normalization (that was the reason for the factor $i/2\pi$ in (5.5)) the classes $c_k(E)$ are in fact cohomology classes with integer coefficients

$$c_k(E) \in H^{2k}(B, \mathbb{Z}), \tag{5.16}$$

and that they coincide with the Chern classes as defined axiomatically (e. g., in Hirzebruch, or Husemoller). In (5.15) we have set $c_0(E) = 1$. Here are the Hirzebruch axioms for Chern classes:

I. For every U(q)-vector bundle over an "admissible space" X, and every integer $i \geq 0$ there is a Chern class $c_i(E) \in H^{2i}(X, \mathbb{Z})$, $c_0 = 1$.

II. Naturality: $c(f^*E) = f^*c(E)$, where $c(E) = \sum c_i(E)$.

III. If $E_1, \ldots, E_q$ are U(1)-bundles over X, then

$$c(E_1 \oplus \cdots \oplus E_q) = c(E_1) \ldots c(E_q).$$

(These axioms are formulated in fact for continuous bundles, wehreas we have considered smooth bundles; an "admissible space" is a locally compact space which is the union of a countable number of compact subsets, and finite dimensional; all paracompact differentiable manifolds are admissible, in this sense.)

Furthermore, it can be shown that the Chern classes of isomorphic bundles coincide.

If E is a trivial bundle the Chern classes are all equal to zero. Thus, we may consider the characteristic classes as a measure of the deviation of a bundle from triviality.

5.2. Pontryagin classes $p_i(X) \in H^{4i}(X, \mathbb{Z})$ of a differentiable manifold are the Pontryagin classes of its tangent bundle. In general, Pontryagin classes are defined like the Chern classes, but of real vector bundles, which have the orthogonal group O(q) as structure group. In turn these can be defined in terms of the Chern classes by embedding U(q) in O(2q) and O(q) in U(q). One thus arrives at cohomology classes of degree 4i

$$p_i(\xi) = (-1)^i c_{2i}(\psi(\xi)), \quad c_{2i+1}(\psi(\xi)) = 0, \qquad (5.17)$$

where ξ is an O(q)-bundle and ψ is a bundle map, coming from the embedding of O(q) in U(q). For details we again refer the reader to either Hirzebruch or Husemoller.

Remark. A number of physics papers have recently discussed the role of integers associated to characteristic classes and, following a practice introduced by Belavin et al.[4], referred to them as Pontryagin numbers or Pontraygin classes. This practice is not quite correct, since the gauge group under consideration was SU(2), and only a special identification with an O(3)-subgroup of SO(4) led to that conclusion. In general, the classes associated to solutions of gauge theories are Chern classes [38], and the integers should be called Chern numbers.

5.3. Homotopy Classes and Homotopy Groups. Integrality Theorems.

In this section we treat two seemingly unrelated topics, which appear together in the applications to gauge theory. It turns out (cf. Section 6.2) that gauges belonging to the same homotopy class are characterized by the same "index", an integral or half-integral valued integral of the top characteristic class of the curvature on a four-manifold. We first summarize pertinent facts about homotopy and then list some integrality theorems.

5.3.1. Homotopies and homotopy classes. We say that two maps (all maps will henceforth be assumed to be continuous) f_0 and f_1 of the topological space X to the topological space Y are homotopic if they are continuously deformable into each other, i. e., if there is a map F: $X \times I \to Y$ ($I = [0, 1]$) such that $F(x, 0) = f_0(x)$ and $F(x, 1) = f_1(x)$, for all $x \in X$. The one-parameter family of maps $f_t = F(\cdot, t)$ is called a homotopy from f_0 to f_1. If two maps f and g are homotopic we write $f \simeq g$. It is easy to see that homotopy is an equivalence relation between maps in the family of all maps from X to Y. The equivalence classes under homotopy are called homotopy classes and the set of all homotopy classes of maps from X to Y is usually denoted by $[X,Y]$. (The reflexivity and symmetry ot homotopy are obvious; transitivity is obtained by defining the composite homotopy $H = F \cdot G$, where f g:X Y with F as homotopy, g h:X Y with G as homotopy and f h with H as homotopy, where

$$H(x, t) = \begin{cases} F(x, 2t), & 0 \le t \le \frac{1}{2} \\ G(x, 2t-1), & \frac{1}{2} \le t \le 1.) \end{cases} \tag{5.18}$$

The identity map $Id_X = 1_X$ is an identity in the family of homotopies of X into itself. A map f: X Y is called a homotopy equivalence with homotopy inverse $g: Y \to X$ if $f \circ g \simeq 1_X$ and $g \circ f \simeq 1_Y$. If f is a homotopy equivalence the spaces X and Y have the same homotopy type and we write $X \simeq Y$. One can show that homotopy of maps is preserved under continuous mappings and that $X \simeq Y$ is an equivalence relation.

A map is said to be homotopic to zero (or null-homotopic) if it is homotopic to the constant map (the map mapping all the points of X into a fixed point of Y). A topological space X is contractible if the map 1_X is homotopic to some constant map of X to itself. Such a homotopy is called a contraction. Any two maps of an arbitrary space to a contractible space are homotopic.

As a familiar example, consider a set in the two-dimensional plane (e. g., the unit disc, or a square) with a hole in it. Then the paths joining two points A and B can be classified into homotopy classes, each class being characterized by the number of times the path winds around the hole. Thus there is a bijection between the homotopy classes of paths and the integers (a negative "winding number" corresponds to winding in the opposite direction, zero to a path from A to B which does not wind around the hole). If we consider the homotopies of closed loops with a fixed base-point, they fall into similar classes. But a loop is a map from the circle, or one-sphere, S^1 into our set. Thus, the homotopy classes of loops in this example are the homotopy classes $[S^1, D]$. These homotopy classes have an obvious group structure, corresponding to the composition of loops, they define the "first homotopy group", or the fundamental group $\pi_1(D)$ of the disc D. It is clear that in the example above of a disc with one hole $\pi_1(D)$ is isomorphic to the group $\mathbb{Z}$ of integers.

5.3.2. Homotopy groups. Generalizing the above example we define the n-th homotopy group of a space X coincides with $[S^n, X]$, the homotopy classes of maps from the n-sphere S^n to X. The group structure is derived from the homotopy properties of the n-sphere S^n. For $n \geq 2$ this group, which will be denoted by $\pi_n(X)$, is abelian. For $n = 1$ we obtain the fundamental group $\pi_1(X)$ which measures the connectivity of the space X . If $\pi_1(X) = 0$ the space is simply connected, i. e., every loop in it is contractible to a point. Other known examples of homotopy groups are : $\pi_3(S^2) = \mathbb{Z}$, $\pi_n(S^n) = \mathbb{Z}$ (here the integers can be interpreted as winding numbers, or as the degree of the map of S^n onto S^n: how many **points** of S^n are taken into one point). Some of the homotopy groups of the classical groups are (cf. Husemoller, sec. 12): $\pi_0(O(n)) = Z_2$, $n \geq 1$; $\pi_0(SO(n)) = \pi_0(U(n)) = \pi_0(SU(n)) = \pi_0(Sp(n)) = 0$, $n \geq 1$. This corresponds to the fact that the groups $O(n)$ consist of two disconnected pieces with positive and negative determinants, whereas all the other groups are connected. The fundamental group, which measures simple-connectedness is: $\pi_1(O(1)) = \pi_1(SO(1)) = 0$. Since $U(1) = SO(2) = S^1$, $\pi_1(O(2)) = \pi_1(SO(2)) = \pi_1(U(1)) = \pi_1(U(n)) = \mathbb{Z}$. It is well known that in $SO(3)$ there are two homotopy classes of loops, hence $\pi_1(SO(3)) = Z_2$ (the group of integers modulo two) and so is $\pi_1(SO(n)) = \pi_1(O(n)) = Z_2$ for $n \geq 3$. The groups $SU(n)$ and $Sp(n)$ are simply connected, and hence their fundamental groups vanish. The second homotopy groups of all compact Lie groups are zero. $\pi_i(O(2)) = \pi_i(SO(2)) = \pi_i(U(1)) = 0$ for $i > 1$ and $= \mathbb{Z}$ for $i = 1$.

Since the three-sphere S^3 can be considered as the set "at infinity" of the four-dimensional euclidean space $\mathbb{R}^4$, the homotopy groups π_3 will play an important role in gauge theories, where a connection which is asymptotically Maurer-Cartan will be defined by an "asymptotic section" $g(x)$ where x is a point on S^3 "at euclidean infinity" and g is in the gauge group. Since $SU(2) = Sp(1) = S^3$ (as topological spaces), we have $\pi_3(U(2)) = \pi_3(SU(2)) = \pi_3(Sp(1)) = \mathbb{Z}$. Further, $\mathbb{Z} = \pi_3(Sp(k))$ for $k \geq 1$, and $= \pi_3(U(k)) = \pi_3(SU(k))$ for $k \geq 2$. $\pi_3(U(1)) = \pi_3(SU(1)) = 0$. Finally, since the space $SO(4)$ is homeomorphic to $S^3 \times SO(3)$, $\pi_3(SO(4)) = \pi_3(S^3) \oplus \pi_3(SO(3)) = \mathbb{Z} \oplus \mathbb{Z}$.

A space X is said to be n-connected if the homotopy groups $\pi_i(X)$ are zero for $i \leqq n$. The Hurewicz isomorphism theorem asserts that if X is $(n - 1)$-connected for $n \geqq 2$, then there is an isomorphism between the n-th homotopy group $\pi_n(X)$ and the n-th homology group $H_n(X)$, the latter can be calculated more easily.

Some physical applications of homotopy can be found in ref.[11]

5.3.2. Integrality theorems. There are quite a number of theorems of this type with which physicists are familiar: the contour integral of an analytic function along a path surrounding a pole is an integral multiple of the residue (the integral is assumed divided by $2\pi i$) the integer being the "homotopy class" of the path, as defined in the preceding section. Similarly, the circulation of the magnetic field (or of a velocity field in two-dimensional fluid dynamics) is an integral multiple of the appropriately normalized current (or vorticity), the integer being again a "winding number". Perhaps one of the least recognized integrality theorems of this kind is the Bohr quantization rule: the integral of pdq (divided by $2\pi\hbar$) is an integer (this interpretation of the quantization rule as a characteristic class is due to Arnol'd).

Differential and algebraic geometry abounds in such integrality theorems, and we describe some of these, in the hope that their analogs in the bundles used in gauge theories may turn out to be of use in physics (in addition to the ones discussed below in Section 6.2).

We have already mentioned Hopf's theorem, according to which the number of zeroes of a smooth vector field on a compact orientable manifold is equal to its Euler characteristic. The latter is also equal to the integral of the top characteristic class. The same is true for compact complex manifolds, where the integral of the top Chern class is equal to the Euler-Poincaré characteristic of the manifold.

The classical theorem in this category is the Gauss-Bonnet-Chern formula: Let D be a domain in a two-dimensional Riemannian manifold with piecewise smooth boundary ∂D. Then the Euler characteristic of the domain is given by the formula:

$$2\pi\chi(D) = \sum_i (\pi - \alpha_i) + \int_{\partial D} \frac{ds}{\rho_g} + \iint_D K dA \qquad (5.19)$$

where $\chi(D)$ is the Euler characteristic (an integer), the sum in the right-hand side is the sum over the exterior angles at the corners of of the boundary, ρ_g is the geodesic curvature of the boundary, and K is the Gaussian (two-dimensional) curvature. Consequently, one may interpret this formula as expressiong the total curvature (0-dimensional plus 1-dimensional plus 2-dimensional), appropriately normalized, as an integer. For a 2-sphere this yields the Euler characteristic 2.

Another integrality theorem, due to Hirzebruch, shows that the signature of a 4-dimensional manifold (also called its index) is equal to its L-genus (these are integers with a relatively complicated definition, cf. Hirzebruch) and is in turn equal to the integral of the first Pontryagin class $p_1(M)$ of the manifold over the compact manifold M, divided by 3. The fact that this Pontryagin number is divisible by 3, has not yet been exploited by physicists (which is surprising, in view of the role played by triplets in particle physics).

One of the most important integtality theorems for compact manifolds is the Atiyah-Singer index theorem (cf., e. g., Hirzebruch, Palais). This theorem relates two indices associated to a differential compact manifold: the analytical index, defined as an integer associated to an elliptic differential operator , and the topological index, defined in terms of characteristic classes of the manifold. The remarkable result of Atiyah and Singer states that these integers are equal. T obe more precise, consider a differential operator D from sections of a vector bundle E over the manifold X to sections of a vector bundle F on X. One can associate to D its symbol (essentially the highest-order terms) which is a homomorphism of the pullbacks of E and F to the unit-ball bundle B(X) of X. D is called elliptic if the symbol is an isomorphism. (Then E and F have the same fiber dimension.) One can define an adjoint D* for D, and two finite-dimensional spaces ker D and coker D (defined like the kernel and cokernel for Fredholm integral operators). The analytic index is then

$$\begin{aligned} i_a(D) &= \dim \ker D - \dim \operatorname{coker} D \\ &= \dim \ker D - \dim \ker D^*, \end{aligned} \tag{5.20}$$

obviously an integer.

The topological index $i_t(D)$ is obtained in a slightly more complicated manner. First one must define the Chern character of D, ch(D) a rational cohomology class depending on the symbol of D. Next one defines the Todd class td(X), of the complexified cotangent bundle of X. The definition is in terms of formal power series and Pontryagin classes too complicated to be described here (cf. the literture). Then the topological index is the value of the cohomology class ch(D)· td(X) on the fundamental cycle of X:

$$i_t(D) = (\mathrm{ch}(D)\cdot \mathrm{td}(X))[X]. \tag{5.21}$$

The Atiyah-Singer index theorem proves the equality of the two indices (this was conjectured earlier by Gel'fand, and checked in special cases).

The index theorem has led to a stormy development of other index theorems, and their potential importance for physics cannot be overestimated. One recent result based on the index theorem is the classification of "instantons" by Atiyah, Hitchin and Singer [65], described in Section 6.2.1, and forthcoming work by Atiyah and Ward, and others.

Characteristic classes and integrality theorems are also bound to play an important role in general relativity. It is striking to note how many important results are particularly true for bundles over 4-manifolds!

I hope to have aroused the curiosity of the reader, to explore these results in more detail. It is time to leave the area of mathematics and give an overview of how these beautiful results might fit into the physical picture.

6. GAUGE FIELDS AND CONNECTIONS

6.0. Introduction. After this long preparation the contents of this chapter amy seem anticlimactic to the reader expecting a complete, systematic treatment of the physics of gauge field theory. Unfortunately it was technically impossible to cover the vast number of applications of differential geometry to gauge theories, or to reformulate in geometric language results which are proliferating through the physics literature. It is the intention of this chapter to sketch some of the most obvious applications and to list a number of curiosities and conjectures, as food for thought of the more adventurous reader. The reader should by now be well equipped to venture out on his own and reformulate some old results in the present language, or obtain some new results.

We will restrict our attention to the following topics: a formulation of classical gauge theories in terms of connections, the meaning of "classical solutions" and of their topological quantum numbers, in particular, the relation to Chern classes, quantization problems, both in the context of a Wightman-type quantum field theory and a description in terms of a Feynman path integral (the latter has been widely used for gauge fields, cf., e. g., the review of Abers and Lee[1]) and a miscellany of observations and conjectures in this connection.

We will not treat here, but hope to return in the future to an analysis of, the various renormalization schemes proposed for gauge theories, as well as a detailed discussion of the Higgs-Kibble mechanism [23, 25, 30] for symmetry breaking. The latter could be given an interpretation in fiber-bundle theory by extending the vector bundles used in gauge theory to "affine bundles", but I feel that this discussion would be premature before we have a better understanding of the quantization problem. Finally, there will only be hints to a very interesting development in gauge theory, namely the theory of gauge fields on a lattice, which has attracted a good deal of attention in recent months (cf., e. g., [7, 21, 39, 34], as well as earlier and unpublished work of K. Wilson). Although at a first glance this approach seems diametrically opposed to differential geometry, and the inherent assumption of smoothness, some of the topological aspects of the two approaches are closely related. In particular, the cohomological properties of gauge fields, which in the lattice approach are "exponentials of connections", or elements of the "holonomy groupoid" (the object one obtains when one replaces the loops used in the construction of the holonomy group by paths; obviously two elements of the holonomy groupoid can be composed only if the beginning of the path of one of them coincides with the end of the path of the other) promise to play an important role in future developments, and in particular, in the path-integral approach to quantization. We hope to return to this topic elsewhere.

6.1. Classical Gauge Fields and Principal Connections. In this section we summarize in a more systematic way many of the results of Section 2.3. It would be pretentious to call the discussion axiomatic. Rather we list a series of assumptions, without testing their independence or consistency. This should be considered as a preliminary to quantization.

1. The base manifold of both the principal fibrations and vector bundles in gauge field theories will be Minkowski space M^4 or the underlying Euclidean space $\mathbb{R}^4$. Since all bundles over $\mathbb{R}^4$ are trivializable, the particular trivialization will amount to a choice of a global gauge. As we shall see on examples, some interesting gauges lead to either singularities at isolated points (in that case the base space will reduce to $\mathbb{R}^4$ with a point or a ball around that point removed), or, even if there is no singularity, to the necessity of using more than one bundle chart. In the latter case some interesting connectivity properties may arise, described by nontrivial characteristic

classes, or by certain degrees of maps ("winding numbers" or similar integers) from the structure group into the sphere S^3 of all the points at infinity of $\mathbb{R}^4$. Although the discussion of $\mathbb{R}^4$ based bundles may seem unphysical, it is at present the only formulation which leads to interesting results (if one does not count lattice gauge theories). In the future one must investigate what restrictions the existence of the light-cone structure in M^4 imposes on gauge theories (e. g., the only curves which should be lifted to the bundle space should be space-like curves; there may be some interesting restrictions on the holonomy groups, etc.) Another generalization awaiting further analysis is to Riemannian, or pseudo-Riemannian manifolds, to take into account background gravitational fields.

2. Classical particle fields are to be considered as (local) sections of a vector bundle over one of the base spaces discussed above. These vector bundles are usually associated via a representation to a gauge group.

3. A gauge group is a principal fibration over the base space, with structure group G, a compact Lie group. The group is not necessarily semisimple and sometimes one may even abandon the compactness assumption (when one wishes to include the Lorentz group as part of the structure group). A gauge is a global section of the principal fibration, and thus it determines a particular trivialization of the bundle, over , e, g., $\mathbb{R}^4 - B$, where B is a ball around the origin. We represent such a section by a function $x \mapsto g(x)$ (rather than by the graph $(x, g(x))$, in order to simplify notation. A gauge is called trivial if it can be continuously deformed into the identity section. It is asymptotically trivial if outside a sphere of sufficiently large radius it can be continuously deformed into the constant section $x \mapsto e$ (the group identity). Thus, gauges fall into homotopy classes of sections which can asymptotically be deformed continuously into each other. Since asymptotically the "boundary" of $\mathbb{R}^4$ is S^3, we will

naturally be led to classifying gauges by the homotopy group $\pi_3(G)$ of homotopic mappings of S^3 into the structure group of the fibration (cf. Section 5.3 for the definitions). For most of the classical groups of importance in physics $\pi_3(G) = \mathbb{Z}$, the additiove group of integers; however $\pi_3(U(1)) = \pi_3(SU(1)) = 0$, and since SO(4) is homeomorphic to $S^3 \times SO(3)$, $\pi_3(SO(4)) = \mathbb{Z} \oplus \mathbb{Z}$ (cf. Husemoller, Section 12).

4. A gauge field is a connection in the principal fibration P, to which the vector bundle of the particle fields is associated. More precisely, we identify a gauge field with the connection 1-form θ or with its coefficients Y in terms of a local basis of the cotangent bundle of the base manifold (dx^i)

$$\omega = Y_i dx^i; \qquad (6.1)$$

the Y_i are matrix-valued objects. The curvature 2-form Ω is defined as the covariant exterior differential

$$\Omega = d\omega + \tfrac{1}{2}[\omega, \omega]; \qquad (6.2)$$

In a basis (dx^i) the coefficients M_{ik} of the two-form Ω are the Yang-Mills field stregth (matrices), and (6.2) takes on the usual form

$$M_{ik} = \partial_i M_k - \partial_k M_i + \tfrac{1}{2}[Y_i, Y_k] \qquad (6.3)$$

which, upon expansion in terms of a basis T^a of the Lie algebra take on the usual form, involving the structure constants (2.81). The difference in the signs of (6.2) and (4.12) is due to the matrix multiplication convention adopted there; note that (6.2) agrees with (4.28), (4.29).

5. The field equations are: a) the Bianchi identity (analog of the homogeneous Maxwell equation):

$$D\Omega = d\omega + \tfrac{1}{2}[\omega, \Omega] = 0; \qquad (6.4)$$

we leave it as an exercise for the reader to write this equation out in terms of Y_i and M_{ik} and to verify that this indeed yields the Yang-Mills equation.

b) The inhomogeneous Yang-Mills equation is written by analogy with the inhomogeneous Maxwell equation in terms of the current three-form *J (the dual of the current of the particles subject to gauge transformations, current which is conserved in a Lagrangian field theory on account of Noether's theorem; here we do not assume a Lagrangian but get the current conservation from a second Bianchi identity):

$$D{*}\Omega = {*}J. \tag{6.5}$$

The fact that D* is covariant-closed

$$DD^{*} = D{*}J = 0 \tag{6.6}$$

implies that the current is conserved in the sense that

$$D{*}J = d{*}J + \omega\wedge{*}J = 0. \tag{6.7}$$

The last term of this equation means that the gauge potential Y itself carries "charge" and gives rise to a current; thus the particle current is not conserved by itself, but rather only together with the current generated by tha gauge field. This, together with the second term in (6.4) gives rise to the nonlinearities characteristic for gauge fields.

6. We do not discuss the Lagrangian formulation of gauge field theory, which we have briefly sketched in Section 2.3. We only note that the appropriate term in the action describing the "free" gauge field is the integral of the four-form formed out of the curvature:

$$A_{YM} = \tfrac{1}{4}\int \Omega\wedge{*}\Omega. \tag{6.8}$$

7. When coupled to a particle field ψ transforming under a representation r(G) of the structure group, the particle field is to be treated as a section of a vector bundel associated to P by that representation and in the field equations satisfied by ψ the ordinary derivatives have to be replaced by the covariant derivatives induced by the connection ω in the associated vector bundle:

$$\partial_i \to \nabla_i = \partial_i + iY_i^a T_a \tag{6.9}$$

where T_a are the matrices representing the basis of the Lie algebra g in the representation r(G) and a summation on a is understood over the appropriate range (the number of group parameters). These results are well known and we do not dwell on them here further. The i in (6.9) is due to the fact that, as is usual in physics, the T_a are hermitean.

6.2. Solutions of the Classical "Free" Yang-Mills Equations. Such solutions have attracted a great deal of attention in recent years (cf., e. g., [4, 10, 11, 26, 31, 35, 36, 45, 53] just to name a few authors). Although considering "free" equations for the gauge fields is somewhat contrary to the philosophy of gauge theory, since there is nothing to be subjected to gauge transformations but the connections themselves, the properties discovered for these solutions are so intriguing that they are worth discussing. We do not have the time here for a full-fledged discussion and refer the reader to the excellent lecture notes by Sidney Coleman[11], but only discuss one interesting example.

We consider a gauge theory described by a principal fibration over $\mathbb{R}^4 - B$, where B is a ball (of arbitrary radius) around the origin, and with structure group $G = SU(2)$. It is to be noted that our base space is a homogeneous space of the group $SO(4) \cong SU(2)\times SU(2)$ and that the "pseudoparticle" or "instanton" solution (due to Polyakov et al. and 't Hooft, loc. cit.) is obtained by making use of a special coupling of the "fiber" SU(2) with one of the two SU(2) groups acting on the base space.

We look for a connection in P which is "asymptotically Maurer-Cartan", i. e., which at large distances from the origin is a "pure gauge"

$$\omega = g^{-1}(x)dg(x), \tag{6.10}$$

or, in terms of the Yang-Mills potentials Y_i:

$$Y_i(x) = g^{-1}(x)\partial_i g(x) \ , \ i = 1,\dots,4. \tag{6.11}$$

Here $g(x)$ is the "gauge function", i. e., the mapping $g\colon \mathbb{R}^4 \to G$ which describes a section of P defined far from the origin. The curvature form Ω or the Yang-Mills field strength M_{ij} vanishes where (6.10) and (6.11) hold, but we assume that at finite distances, in particular near the surface of the ball B, the connection Y_i is nontrivial and that there is nonzero curvature M_{ij}. The "instanton" (or "pseudoparticle" connection of 't Hooft and Polyakov is, up to a coupling constant, which we have absorbed into the definition of Y_i, M_{ij} and which is absent from the Y-M equations (6.3),(6.4)[in the physics literature our factor ½ in these two equations is usually replaced by -ig, the -i being due to the fact that the $A_j = \frac{i}{g}Y_j$ are hermitean, and g is a coupling constant characterizing the self-coupling of the Y-M field, and which we have omitted, in order to bring out the geometric nature of the equations):

$$Y_j = i\frac{\eta_{ajk}\tau^a x^k}{r^2 + \lambda^2}, \qquad (6.12)$$

where the coefficient η_{ajk} is a tensorial quantity which couples representations of the structure group SU(2) with represent tions of the invariance group SO(4) of the base space in the following manner:

$$\begin{aligned} \eta_{ajk} &= e_{ajk} \quad (a, j, k = 1, 2, 3), \\ \eta_{a4k} &= -\delta_{ak} \quad (a, k = 1, 2, 3), \\ \eta_{aj4} &= \delta_{aj} \quad (a, j = 1, 2, 3), \\ \eta_{a44} &= 0; \end{aligned} \qquad (6.13)$$

$i\tau^a$ are the generators of the Lie algebra of SU(2), with τ^a the isospin Pauli matrices and e_{ajk} is the Levi-Civita tensor of $\mathbb{R}^3$. In (6.12) $r^2 = (x^1)^2 + (x^2)^2 + (x^3)^2 + (x^4)^2$. The **constant** λ^2 prevents Y_j from becoming singular at the origin (in our setting this is not important, since we have eliminated the origin anyway, but in other contexts one might want to translate the solution (6.12) to other points.

The connection (6.12) can be obtained from the following g(x):

$$g(x) = (r^2 + \lambda^2)^{-1}(x_4 + i\tau_a x^a), \qquad (6.14)$$

where the summation over a is from 1 to 3. It is easily verified that

$$Y_j = \frac{r^2}{r^2 + \lambda^2}\, g^{-1}(x)\partial_j g(x), \qquad (6.15)$$

i. e., for $r \to \infty$ this is a "pure gauge" of the Maurer-Cartan form (6.11). The section g(x) realizes a mapping of the sphere S^3 (all the points at infinity of $\mathbb{R}^4$) onto the group SU(2) which covers SU(2) exactly one time, and cannot be deformed continuously into g(x) = e, the group identity. In general, $[g(x)]^n$ with n a positive or negative integer, will produce a mapping covering SU(2) n times, and for different n these mappings are not homotopic to each other, i. e., belong to different homotopy classes. The integers n, called occasionally the "topological charge" are elements of the homotopy group $\pi_3(SU(2)) = \mathbb{Z}$.

The solution (6.12) has many interesting properties, which have been widely discussed in the literature, but we restrict our attention to only one of them, namely the calculation of the Chern number c_2. Since our base space is four-dimensional, the highest nonvanishing Chern class of the connection Y_j can be $c_2(P)$. In this particular case the only four-form calculated according to the determinant (5.5) will be , in terms of the Yang-Mills field strength matrix M_{jk}

$$c_2(P) = -(16\pi^2)^{-1} M_{jk} M_{lm}\, e^{jklm} d^4x, \tag{6.16}$$

where one factor of ¼ comes from the antisymmetric tensor e^{jklm} (which has been introduced in order to avoid the oriented volume element) and $-(1/4\pi^2)$ gives the normalization of the determinant (5.5).

In this particular example, the integration of the four-form (6.16) over the whole of $\mathbb{R}^4$ is easily carried out, since $c_2(P)$ is closed, and we can replace it by the differential of a three-form and apply Stokes theorem. The calculation is fairly straightforward, and based on the fact that (up to the numerical factor in front)

$$c_2(P) = d\{\omega \wedge (d\omega + \omega \wedge \omega)\}, \tag{6.17}$$

or, in terms of the Yang-Mills potentials A_j (cf.[4, 54]) the integrand is the 4-divergence

$$\partial_i [e^{ijkm} (A_j \partial_k A_m - 2igA_j A_k A_m] . \tag{6.18}$$

Stokes' theorem (in this case, the 4-dimensional version of Gauss' theorem) reduces the integral to a three-dimensional integral over S^3. But on S^3 at infinity the connection is Maurer-Cartan (6.10), and the three-dimensional integral reduces to the Haar integral over SU(2), which is equal to $8\pi^2$. The correct factors then yield the value one.

The "instanton" solution (6.12) exhibits the important feature that its curvature M is self-dual

$$M = *M, \tag{6.19}$$

whereas the "anti-instanton" solution, with Chern number -1 is "anti-selfdual"

$$M = -*M. \tag{6.20}$$

Another classical solution of the Yang-Mills equations has been obtained recently by De Alfaro, Fubini, and Furlan [68], from their analysis of conformally invariant field theories. This solution has the form

$$Y_j = \frac{i}{2}\,\frac{\eta_{ajk}\tau^a x^k}{r^2}, \tag{6.21}$$

with

$$M_{jk} = [\eta_{ajr}\tau^a x^r,\ \eta_{bks}\tau^b x^s]\tfrac{1}{4}r^{-4}. \tag{6.22}$$

Here the notations are the same as above, and we must warn the reader that we have absorbed the coupling constant and a factor of 2 in each of the η matrices into the definitions of the fields. These solutions are, of course, singular at the origin and at infinity, but a conformal transformation allows one to place the singularities at two points at finite distances, so that they can be interpreted as "lumps of one-half the topological charge", which have been called "merons" by Callan, Dashen and Gross [67]. It should be noted that the "Pontryagin index" used in the physical literature differs by a factor of two from the one used by the mathematicians, and as we have said before, it is more appropriate to use the Chern number , since we are dealing with unitary groups as structure groups.

6.2.1. The results of Atiyah, Hitchin, and Singer [65]. Very recently (April-May 1977), I have learned from Prof. I. M. Singer of a study made of self-dual curvatures of bundles on four-manifolds, which have yielded, in particular, the result that for the SU(2) gauge theory in Euclidean 4-space the classical solutions which are self-dual, depend on 8k - 3 parameters, where k is the Pontryagin index (Chern number).

Atiyah, Hitchin, and Singer consider a connection A with curvature F in a fibre bundle with structure group G over $\mathbb{R}^4$, in particular, the cases G = SU(2) and G = SU(n). The connection is assumed to be asymptotically flat at infinity and the action $\|F\|^2 < \infty$. They impose the restriction on A that at infinity it extends to a connection for a bundle over the 4-sphere S^4. The topological type of such a bundle is determined by $\pi_3(G)$, i. e., by an integer k for G = SU(n) (cf. Sec. 5.3). This integer is also equal to the Chern number defined in the preceding section, and is variously called Pontryagin index, or winding number in the physics literature (as we have pointed out, it differs by a factor of 2 from the mathematical terminology).

The curvature F is decomposed into the direct sum $F = F^+ \oplus F^-$, where $F^+ = *F^+$ is self-dual and $F^- = -*F^-$ is anti-self-dual. Then Eq. (6.16) shows that

$$k = \int c_2(P) = (8\pi)^{-2}[\|F^+\|^2 - \|F^-\|^2] \qquad (6.23)$$

(the normalization differs from the one used in (6.16); some arbitrariness enters due to the definition of the coupling constant).

If one looks for minima of $\|F\|^2$ one must have either $F^+ = 0$ or $F^- = 0$, and then , for self-dual solutions $\|F^+\|^2 \geqq 8\pi^2 k$. For k = 0 one obtains the trivial F = 0, for k = 1 we have the instanton (6.12), etc. It seems that Jackiw, Nohl, and Rebbi have constructed solutions depending on 5k + 4 parameters.[70]

Atiyah, Hitchin and Singer have considered the space of all solutions A of the Yang-Mills equations modulo the action of the gauge group and have proved that this so-called <u>space of moduli</u> of self-dual SU(2) connections over S^4 with index k 1 is a manifold of dimension

$$h^1 = 8k - 3. \qquad (6.24)$$

An important ingredient of the proof is the use of the Atiyah-Singer index theorem for elliptic operators on manifolds. The arguments extend to G = SU(n) if the connection is irreducible. The result is then $h^1 = 4nk - n^2 + 1$. The result (6.24) was derived "physically" in.[66]

6.3. Quantum Theory of Connections. In this section we attempt to formulate a quantization scheme for gauge fields which recognizes that they are connections in principal fibrations. The usual problems which appear in the quantization of the electromagnetic field, such as the need for an indefinite metric in the Hilbert space used in the Bleuler-Gupta formalism is a consequence of the fact that the electromagnetic potentials are connection forms which overdetermine the field, whereas the field-strengths are curvature forms, which to some extent under-describe the situation (e. g., do not properly account for the Bohm-Aharonov effect, unless one uses their topological characteristics). We start out with a review of the Wightman (or Gårding-Wightman) axioms for charged fields and then propose a tentative treatment of the connection forms within this framework.

6.3.1. Quantized particle fields. The extension of the Gårding-Wightman axioms [19] to fields which exhibit an "internal" symmetry given by a representation D(G) of a compact symmetry group G is quite straightforward. For simplicity we shall assume that the fields are Lorentz scalars, but the extension to spinors or other higher-order quantities is straightforward too.

It will be convenient to consider the fields as distributions over sections of a trivial vector bundle, and later to make the bundle non-trivial by replacing the constant group action by the action of a principal fibration (gauge group of the second kind).

We therefore consider as test-functions cross sections $f_a(x)$ of a d-dimensional trivial vector bundle over Minkowski space, transforming under the d-dimensional representation D(G) of G. These sections are infinitely differentiable and belong either to the space $\mathcal{S}$ (test functions which together with all their derivatives decrease at infinity faster than any reciprocal power of the euclidean distance to the origin) or to the space $\mathcal{D}$ (infinitely differentiable test functions of compact support. We define the operator-valued distribution

$$\varphi(f) = \sum_a \varphi_a(\bar{f}_a) = \sum_a \int \varphi_a(x)\bar{f}_a(x)d^4x, \tag{6.25}$$

where the integral in the last expression is of heuristic value and where the complex conjugate $\bar{f}_a$ has been introduced for later convenience. The action of the structure group G on the field operator is obtained by "transposing", in the usual manner of distribution theory, the action of the representation D(G) on f_a, and then taking into account the unitarity of the representation D(G):

$$\sum \int D_{ab}(g)\varphi_b(x)\bar{f}_a(x)d^4x = \sum \int \varphi_b \overline{[\bar{D}_{ab}(g) f_a(x)]} d^4x$$

$$= \sum \int \varphi_b(x) [D(g^{-1})]_{ba}\bar{f}_a(x)d^4x. \tag{6.26}$$

Hence, we adopt as a definition of the action of G on the operator-valued distribution $\varphi(f)$:

$$U(g)\varphi(f)U(g)^{-1} = \varphi(D^T(g^{-1})f), \tag{6.27}$$

where D^T denotes the transpose of the matrix D.

Next we replace the group G, acting identically at all points of the underlying Minkowski space M, by a principal fibration P with G as structure group and M as base space. Then the "gauge transformation of the second kind" (i. e., the point dependent gauge transformation) is defined by a (possibly local) section g(x) of the principal bundle P. Any section f of the vector bundle associated to P by the representation D^T can be viewed as the "pullback" of a section s(x) of P via the relation

$$f(x) = s(x)\zeta(s(x)), \tag{6.28}$$

where ζ is a C^∞-map of a neighborhood of $s(x) = p_x \in P_x$ in the fiber over x of P into the fiber E_x of the associated vector bundle (i. e., a map from the group into the vector space carrying the representation D, at least locally), and s(x) on the left of the right-hand side of (6.28) is the action on that vector (for simplicity we have suppressed the extra D which should appear in this formula).

In particular, support properties of f(x) will carry over to the mapping ζ (since it does not make sense to talk of vanishing of s(x)).

For simplicity, since we will have problems of multiplication, we assume the ttest-sections f(x) to be of compact support, i. e., in $\mathcal{D}(M, E)$, where E denotes the associated vector bundle.

Since the group G acts on sections s(x) of P by right multiplication, we can represent the action of a section g(x) of P (a local gauge transformation) on a section f(x) of E by a "point-dependent" representation $D^T[g^{-1}(x)]$ (valid at least locally):

$$(g\cdot f)(x) = D^T(g^{-1}(x))f(x) = g(x)s(x)D^T(g^{-1}(x))\zeta(g\cdot s(x)), \quad (6.29)$$

where products and inverses of sections of P are defined pointwise.

We now assume the following action of a section g(x) of P on the operator-valued distribution $\varphi(f)$:

$$U[g]\varphi(f)U[g]^{-1} = \varphi(g^{-1}(x)\cdot f(x)) .= g\cdot\varphi(f). \quad (6.30)$$

The notation in the left-hand side is highly symbolic, since we do not know whether a unitary operator U[g] which implements the action exists even for sections which deviate from the identity section only in a small neighborhood of x; if the base space is the whole of Minkowski space the fibration P is trivializable and then (6.27) implies (6.30). The situation in nontrivial bundles is complicated by the nonexistence of global sections and needs further careful analysis.

Since we have assumed f of compact support, the middle term of (6.30) is well defined. In a certain sense U[g] should be considered a functional of the section g(x), fact which is symbolized by the use of square brackets.

Let us return to the situation where the group acts identically on all fibers (i. e., we have gauge transformations of the first kind). We then assume that the field $\varphi(f)$ satisfies the Gårding-Wightman axioms for "charged" fields, i. e.,

1. There exists a Hilbert space H carrying a continuous representation (or projective representation) $U(P_+^\uparrow)$ of the proper orthochronous Poincaré group or of its covering group. Suppose there is a unique vector $\Omega \in H$ invariant under this representation and that the spectrum

of the generators P_k of space-time translations, $U(a, 1) = \exp(iP_k a^k)$, is in the forward light-cone. If we are dealing with tensor fields, we consider the Poincaré group itself, if spinor fields are involved, we have to use the covering group. Furthermore, if one considers non-orthochronous trnasformations, the representation is anti-unitary. Furthermore, there is a representation U(G) of the "internal symmetry group" G, which commutes with the representation $U(\mathcal{P})$.

2. The operators $\varphi(f)$ and $\varphi(f)^*$, where f is a tensor or spinor under Lorentz transformations and transforms according to the representation D(G) of the gauge group have a common dense domain $D \in \mathcal{H}$, containing the vacuum vector Ω, and is invariant under the representations $U(\mathcal{P})$ and U(G). Further, these field operators are weakly continuous, i. e., their matrix elements between vectors of D are distributions on the appropriate test-function space.

3. The operator-valued distributions satisfy the transformation law (6.27) under U(G) and a similar law under inhomogeneous Lorentz transformations. For simplicity we will pretend that the fields are scalars under Lorentz transformations.

4. Local commutativity: if the supports of the test-functions f_1 and f_2 are space-like separated, i. e., if $f_{1a}(x)f_{2b}(y) = 0$ for all x and y such that $(x - y) \geqq 0$, then if f_1 and f_2 are tensors under the Lorentz group

$$[\varphi(f_1), \varphi(f_2)] = 0, \tag{6.31}$$

whereas if f_1 and f_2 are spinors

$$\{\varphi(f_1), \varphi(f_2)\} = 0, \tag{6.32}$$

where the curly bracket denotes anticommutators.

Spinor fields and charged fields are "unobservable" in the sense that they give rise to superselection rules in the Hilbert space of the theory. For spinors, the two-valuedness of the representation of the rotation subgroup produces the univalence superselection rule (no superposition of states of integer-spin and half-integer spin particles).

If the gauge group is $U(1) = T^1$, the corresponding superselection rule is the usual charge superselection. If the gauge is an n-dimensional torus T^n, we have n conserved charges (such as electric charge, baryon number, lepton number, etc.). The observables of the theory are in the vacuum sector, and are obtained by "averaging" over the compact gauge group. In the case of nonabelian gauge groups the algebraic approach (the von Neumann algebra generated by the polynomials of the smeared fields) leads to the conclusion that superselection is equivalent in a certain sense to parastatistics (cf. [13] for details).

In addition to the above axioms, which remain essentially the same in the fiber-bundle approach, quantized fields also obey field equations which are partial differential equations to be understood in a distribution sense. Thus, for scalar fields, the test functions $f(x)$ are assumed to be solutions of the Klein-Gordon equation (or of the Duffin-Kemmer system we have discussed in Sec. 2.3). This is modified, if one goes over from a gauge group of the first kind to a gauge group of the second kind, i. e., to the action of a principal fibration on the vector bundle of the particle fields.

The transition to the bundle situation is most easily achieved in terms of a <u>moving frame</u> in the vector bundle E associated to the principal fibration P. A <u>moving frame</u> is a set of linearly independent (local) sections

$$e_a(x) = D^T(g^{-1}(x))_{ab}e_b, \tag{6.33}$$

where e_b is a fixed basis for the framing representation of the associated vector bundle for which the test-functions f are sections and $g(x)$ is a local section (gauge) of the principal fibration.

We then replace the derivatives ∂_i occuring in the partial differential equations obeyed by the test-functions f by the appropriate gauge-covariant derivatives ∇_i, which in turn, in terms of a moving frame, can be expressed in terms of point-dependent connection matrices Y_i. The covariant derivative of the field $\varphi(f)$ is then defined in the usual distribution-theoretic manner

$$(\nabla_i \varphi)(f) = \varphi(-\nabla_i f) = (\partial_i \varphi)(f) + (Y_i \cdot \varphi)(f), \qquad (6.34)$$

where the matrix Y_i in the last expression is understood as acting on the appropriate indices in a heuristic representation of the type of Eq. (6.25).

6.3.2. Quantized connection forms? The question arises of how to interpret the connection form Y in the expression (6.34), and the curvature form M obtained from it by covariant differentiation. As long as we use classical gauge theory and the lagrangian formalism as a guide, we would be tempted to treat Y_i and M_{ik} as field operators. But already the analysis of the electromagnetic field leads here to the added complication of "indefinite metric quantization" a la Bleuler and Gupta. We return to this point below, but first discuss briefly other alternatives.

One way of interpreting Eq. (6.34) is to consider Y (and M) as operator valued multipliers of the distributions describing the field φ. That means that the symbol $(Y_i\varphi)$ is again an operator-valued distribution of the same kind as φ. For numerical distributions on various test-function spaces the theory of multipliers is well developed: there are classes of functions which can multiply a distribution leaving it a distribution. In our situation the "operator" Y is a representation of the Lie algebra g by unbounded operators, and we have to impose the stringent requirement that its domain be in the domain D of the fields. We shall return to a detailed study of these questions elsewhere, but would like to call the probelm to the reader's attention.

Another interpretation of the connection form Y in a quantized theory is an "operator-morphism" which maps the operator φ into the new operator $Y \cdot \varphi$. This point of view would be advantageous in the context of the von Neumann algebra generated by the fields, the so-called field algebra $\mathcal{F}$, which can be obtained from the observable algebra $\mathcal{A}$ by the well known extension approach of [13]. This problem will be discussed in a future publication.

The Bleuler-Gupta approach to quantizing connection forms was extended to the nonabelian case, and is described in my Bonn lecture [37]. It is not known whether the postulates made there are compatible and whether there exists a nontrivial example of gauge theories satisfying these generalized axioms. The perturbation approach to gauge theories, the success of quantum electrodynamics, and the recent results on gauge theories on lattices are encouraging.

On the other hand, there does not exist to date a rigorous theory of symmetry breakdown (nonuniqueness of the vacuum), in the fiber-bundle approach to gauge theories. In the theory of symmetry breakdown an essential role may be played by the holonomy group of the connection. It is usually assumed that when symmetry breakdown occurs, i. e., the vacuum vector is not unique and not invariant under the gauge group, that there is a residual symmetry group. It is easy to see that the residual symmetry group must be the holonomy group of the connection. For some nonabelian groups, the residual group may turn out to be abelian. But a connection with an abelian holonomy group will have an abelian curvature form (by the Ambrose-Singer theorem), which will satisfy Maxwell rather than Yang-Mills equations[31].

Another context in which the symmetry group must be abelian is the superselection theory of Doplicher, Haag, and Roberts[13], where the field algebra is required to satisfy "duality"

$$[F(\mathcal{O})]' = F(\mathcal{O}') \qquad (6.35)$$

where F is the field algebra associated to the open double cone $\mathcal{O}$, F' denotes the commutant of the von Neumann algebra and $\mathcal{O}'$ denotes the region of spacetime which is spacelike to the double cone $\mathcal{O}$. This property was known for free fields from early work of Araki, and has been recently proved to hold for field theories satisfying the Wightman axioms by Bisognano and Wichmann. It is therefore tempting to assume it as a property of field theory. But then the symmetry group must be abelian, and the holonomy group is a natural candidate for this role. We are thus led to a tempting conjecture: whereas the internal symmetry of a field may be described by a principal fibration with anonabelian compact structure group, the holonomy group is abelian. Space-time translations commute only with the holonomy. Nonabelian curvature is confined to the inside of particles, which would account for confinement of quarks and gluons. Details of this picture still need to be worked out.

One last remark on quantization. We have seen that classical solutions of the Yang-Mills equations are elements of characteristic classes, at least in four-dimensional Euclidean space. What is the role of these classes, and the associated integrality theorems in the quantization of gauge fields? It is appropriate to point out that the old Bohr quanization rule also made use of a characteristic class: the loop integral of the 1-form pdq is an integral multiple of 1 (if Planck's constant is set equal to unity). The role of this characteristic class in quantization and its connection to "wave-front" dynamics was pointed out by Arnol'd (cf. the Appendix to his book, quoted in the bibliography and an article quoted there). Could it be that the Chern or Pontryagin numbers might play a similar role in the quantization of the connections? More precisely, what is the correct quantization of the connection form and its curvature, so that the integers or half-integers encountered in the characteristic classes for classical solutions are eigenvalues of certain operators? The recent results of Atiyah and Singer may shed some light on this problem.

6.4. Feynman Path Integrals. We limit ourselves to some brief remarks on this most popular method of quantization of gauge fields. It has been extremely successful in discussing the Weinberg-Salam model of weak and electromagnetic interactions, renormalization problems in perturbation theory, the lattice approach to gauge fields and has led to the discovery of the integrality theorems mentioned above.

In the path-integral approach to quantum mechanics (cf. Feynman and Hibbs, or Abers and Lee [1]) for systems with classical action

$$S_{cl} = \int_{t_0}^{t_1} L(Q, \dot{Q})dt \tag{6.36}$$

the vacuum to vacuum transition amplitude is expressed as a "functional integral" or path integral

$$\langle Q_1, t_1 | Q_0, t_0 \rangle = \int \mathcal{D}Q(t)\exp[(i/\hbar)S_{cl}] \tag{6.37}$$

where the integration element is a "measure" on the space of all classical paths.

Replacing the action (6.36) by the action of a field theory, and adding a "source term" to the Lagrangian, one can obtain a generating functional for the vacuum expectation values of time-ordered products of field operators (cf. the above references, for details).

Thus, for a scalar field theory the generating functional has the form

$$W[J] = \int \mathcal{D}\phi \exp[i\int (\mathcal{L}(x) + J(x)\phi(x))d^4x], \tag{6.38}$$

so that the connected Green's functions can be obtained by "logarithmic functional differentiation" from W[J]:

$$G_c(x_1,\ldots,x_n) = (-i)^n \frac{\delta^n \ln W[J]}{\delta J(x_1)\ldots J(x_n)} . \tag{6.39}$$

When one attempts to extend this method to gauge fields, i. e., to connections, one runs into the following difficulties, discovered by Feynman[17] and De Witt[12] and partially resolved by Faddeev and Popov[15]. Owing to the freedom of performing gauge transformations, the integration over "paths" in the space of gauge fields is not well defined, but rather over a subset of paths, obtainable from each other by gauge transformations. In order to make a path integral meaningful, Faddeev and Popov have proposed to introduce a "delta-functional" taking into account the gauge condition (e. g., divA = 0, in the Coulomb gauge), and to eliminate the redundant integration over the "volume" of the orbits of the gauge group.

If one attempts to translate the method into the language of principal fibrations and connections one comes up with the interesting mathematical problem of defining a path integration over equivalence classes of connections, which takes into account the ambiguity in the choice of gauge (i. e., the changes of charts in the principal fibration). So far, little attention has been devoted to the geometric aspects of the Faddeev-Popov prescription, the meaning of the gauge-fixing terms, etc.

This problem is being actively pursued in the Euclidean lattice theory framework[21, 42] and we hope to return to it elsewhere. Here we only remark that the search for classes of gauges over which the functional integration is to be carried out has led to the discovery of the homotopy classes of "vacua" discussed earlier and the importance of the concept of characteristic classes for gauge theory. For further details we refer the reader to the literature.

6.5. Remarks and Conjectures. We close these lecture notes with a few remarks and conjectures, which require further investigation.

We first note that our discussion of quantization of connection forms, both in a Wightman type approach, and in a Feynman path integral approach was rather vague and incomplete. This is in part due to the unsatisfactory state of affairs, and in part to the fact that I have not been able to solve some of the problems I have attacked. We have not discussed at all the problem of symmetry breaking, Higgs bosons, etc., which has been widely discussed in the physical literature. One conjecture which I would like to mention is that a proper treatment of the Higgs mechanism will require the introduction of "affine" vector bundles, since the existence of fields with nonvainshing expectation values requires shifting some vector variables.

Another important remark is that the discussion of "pure gauge fields" somewhat violates the spirit of Weyl's gauge principle, and that in general, one should always consider the gauge fields together with the particle fields to which they are coupled. The only exception may be the electromagnetic field, where a classical counterpart does indeed exist, and where in compact regions of spacetime one may consider free electromagnetic fields. This is in agreement with the empirical fact that no particles corresponding to nonabelian gauge fields (intermediate vector bosons, gluons) have been observed so far, and they may well not be observed even when the available energies will be considerably higher than thos presently available.

It is imperative to investigate in much more detail the role of characteristic classes (and other cohomology properties) of the connections describing gauge fields, and their connection with the quantization scheme, be it in terms of "quantized connections" or in terms of path integrals. The recently discovered classification and enumeration of classical solutions to the Yang-Mills equations and their generalizations[65], the discovery of "merons" [67, 68], and the analysis of cohomologies of lattice gauge theories, are of the utmost importance, and are under active investigation.

Finally, let me only note that there is some indication that gauge properties may play an important role also in magnetohydrodynamics , where recently redicovered integral invariants involving the vector potential may lead to important new developments both of principle and applications.

BIBILIOGRAPHY

A. BOOKS (quoted by name of author(s) and volume)

Arnol'd, V. I.: Matematicheskie metody klassicheskoi mekhaniki (Mathematical methods of classical mechanics), Nauka, Moscow, 1974.

Bourbaki, N: Variétés différentielles et analytiques, Hermann, Paris, §§ 1 - 7, 1967, §§ 8 - 15, 1971.

Chern, S-S.: The Geometry of Characteristic Classes, in Proc. of the 13-th Biennial Seminar, Canadian Math. Congress, Montreal, pp. 1 - 40.

Chevalley, C.: Theory of Lie Groups, vol.I, Priceton U. Press, Princeton, 1946.

De Rham, G.: Variétés différentiables, Hermann, Paris, 1955.

Dieudonné, J.: Treatise on Analysis, Academic Press, N. Y., vol III, 1972, vol. IV, 1974 (vols. V, VI in French).

Greub, W., S. Halperin, and R. Vanstone: Connections, Curvature, and Cohomology, Academic Press, N. Y., Vol. I., 1972, Vol. II, 1973, Vol. III, 1974.

Hawking, S. W., and G. F. R. Ellis: The Large-Scale Structure of the Universe, Cambridge University Press, 1973.

Hermann, R.: Vector Bundles in Mathematical Physics, Benjamin, N.Y. and a large number of other books by the same author.

Hirsch, M.: Differential Topology, Springer Verlag, N. Y., 1976.

Hirzebruch, F.: Topological Methods in Algebraic Geometry, 3-rd ed. Springer Verlag, Berlin, 1966.

Husemoller, D.: Fibre Bundles, 2-nd ed., Springer Verlag, N. Y. 1975.

Kobayasi, S., and K. Nomizu, Foundations of Differential Geometry, Wiley, .N. Y., Vol. I, 1963, Vol, II, 1969.

Landau, L. D., and E. M. Lifshitz: Teoriya polya (Classical Field Theory) 6-th ed., Nauka, Moscow, 1973; Engl. Transl. Pergamon Press.

Lichnerowicz, A.: Theorie globale des connexions et des groupes d'holonomie, Ed. Cremonese, Roma, 1955.

Lightman, A. P., W. H. Press, R. H. Price, and S. A. Teukolsky:Problem Book in Relativity and Gravitation, Princeton U. P.,1975.

Mayer, M. E.: Cîmpuri cuantice si particule elementare (Quantized fields and elementary particles), Ed. Tehnica, Bucharest, 1969.

Misner, C. W., K. S. Thorne, and J. A. Wheeler, Gravitation, Freeman, San Francisco, 1973.

Pauli, W.: Die allgemeinen Prinzipien der Quantenmechanik, Handb. d. Physik, 2-nd ed., Bd. V, T. 1, Springer Verlag, Berlin, 1958.

Spivak, M.:A Comprehensive Introduction to Differential Geometry, 3 vols. , Publish or Perish, Waltham, 1971. Also: Calculus on Manifolds, Benjamin, 1969.

Steenrod, N.; The Topology of Fibre Bundles, Princeton Univ. Press, 1951.

Sternberg, S: Lectures on Differential Geometry, Prentice-Hall, Englewood Cliffs, 1963.

Sulanke, R., and P. Wintgen, Differentialgeometrie und Faserbündel, Birkhäuser Verlag, Basel 1972.

Vranceanu, G.: Leçons de geometrie differentielle, 3 vols. Ed. de l'Académie de la R. P. Roumaine, Bucharest/ Gauthier-Villars, Paris, 1964 (also available in German and the Romanian original).

Yang, C. N.: Lecture notes on gauge theories and fiber bundles (approximate title), University of Hawaii, 1975.

Milnor, J. W., and J. D. Stasheff: Characteristic Classes, Princeton Univ. Press, 1974.

B. ARTICLES (including review articles)

[1] E. Abers and B. W. Lee, Gauge Theories, Phys. Reports, 9,No 1, 1 - 141 (1973).

[2] W. Ambrose and I. M. Singer, A Theorem on Holonomy, Trans. Amer. Math. Soc. 75, 428 - 443 (1953).

[3] R. L. Arnowitt and S. I. Fickler, Phys. Rev. 127, 1821 (1962).

[4] A. A. Belavin, A. M. Polyakov, A. S. Schwartz and S. Tyupkin, Pseudoparticle Solutions of the Yang-Mills Equations, Phys. Lett. 59B, 85 (1976).

[5] J. Bernstein, Spontaneous Symmetry Breaking, Gauge Theories, Higgs Mechanism and All That, Rev. Mod. Phys. 46, 1 (1974).

[6] K. Bleuler, Helv. Phys. Acta 23, 567 (1950).

[7] R. Balian, J. Drouffe and C. Itzykson, Gauge Fields on a Lattice, Phys. Rev. D10, 3376 (1974); D11, 2098 (1975).

[8] R. Bott and S. S. Chern, Hermitian Vector Bundles and the Equidistribution of Zeroes of their Holomorphic Sections, Acta Mathem. 114, 71 - 112 (1965).

[9] R. Brout and F. Englert, Phys. Rev. Lett. 13, 321 (1964).

[10] C. Callan, R. F. Dashen and D. J. Gross, The Structure of the Gauge Theory Vacuum, Phys. Lett. 63B, 334 (1976); A Mechanism for Quark Confinement, Princeton Preprint COO-2220-94, 1977.

[11] S. Coleman, Classical Lumps and their Quantum Descendants, Erice Lecture Notes, 1975.

[12] B. S. De Witt. Phys. Rev. 162, 1195, 1239 (1967).

[13] S. Doplicher, R. Haag and J. E. Roberts, Fields, Observables, and Gauge Transformations, I, II, Commun. Math. Phys. 13, 1 -23 (1969); 15, 173 - 200 (1971). Local Observables and Particle Statistics, I, Ibid. 23, 199 - 230 (1971); S. Doplicher and J. E. Roberts, Fields, Statistics and Nonabelian Gauge Groups, Ibid. 28, 331 - 348 (1972). Lectures by S. Doplicher at various conferences.

[14] L. D. Faddeev, Lecture at the Mathematical Physics Conference, Moscow, 1972; published in the Trudy Matem. Inst. im. Steklova (in Russian) 1975.

[15] L. D. Faddeev and V. S. Popov, Feynman Diagrams for the Yang-Mills Field, Phys. Lett. 25B, 29 (1967); Kiev Preprint, 1967; L. D. Faddeev, Teor. Matem. Fiz, 1, 3 (1969) [Theor. Math. Phys. 1,1 (1969)].

[16] E. Fermi, Rev. Mod. Phys. 4, 87 (1932).

[17] R. P. Feynman, Acta Physica Polonica 26, 697 (1963).

[18] P. L. Garcia, Gauge Algebras, Curvature and Symplectic Structure to appear in J, Differ. Geom.; Reducibility of the Symplectic Structure of Classical Fields, Proc. of the Symposium on Differential-Geometrical Methods in Physics, Bonn, 1975, K. Bleuler and A. Reetz, eds., Springer Lecture Notes in Math., 1977.

[19] L. Gårding and A. S. Wightman, Fields as Operator-Valued Distributions in Relativistic Quantum Field Theory, Arkiv för Fysik 28, 129 - 184 (1964).

[20] M. Gell-Mann and S. L. Glashow, Ann. Phys. (N. Y.) 15, 437 (1961).

[21] J. Glimm and A. Jaffe, Quark Trapping for Lattice U(1) Gauge Fields, Harvard Preprint, 1977. Functional Integral Methods in Quantum Field Theory, Cargese 1976 Lectures, to be published.

[22] S. N. Gupta, Proc. Phys. Soc. (Lond.) 61A, 68 (1950).

[23] G. S. Guralnik, C. R. Hagen and T. W. B. Kibble, Phys. Rev. Lett. 13, 585 (1964).

[24] S. W. Hawking, Gravitational Instantons, Phys. Lett. 60A, 81 (1977).

[25] P. W. Higgs, Phys. Rev. Lett. 12, 132 (1964); Phys. Rev. 145, 1156 (1966).

[26] R. Jackiw and C. Rebbi, Vacuum Periodicity in a Yang-Mills Theory, Phys. Rev. Lett. 37, 172 (1976). Conformal Properties of a Yang-Mills Pseudoparticle, Phys. Rev. D 14, 517 (1976).

[27] H. Kerbrat-Lunc, Ann. Inst. H. Poincaré 13A, 295 (1970).

[28] R. Kerner, Ann. Inst. H. Poincaré 9A, 143 (1968).

[29] T. W. B. Kibble, J. Math. Phys. 2, 212 (1961).

[30] T. W. B. Kibble, Phys. Rev. 155, 1554 (1967).

[31] H. G. Loos, Internal Holonomy Groups of Yang-Mills Fields, J. Math. Phys. 8, 2114 - 2124 (1967); Phys. Rev. 188, 2342 (1969).

[32] Lu Qi-keng, Gauge Fields and Connections in Principal Bundles, Chin. J. Phys. 23, 153 - 161 (1975)[Wuli Xuebao 23, 249-263 (1974)]

[33] E. Lubkin, Ann. Phys. (N. Y.) 23, 233 (1963).

[34] V. Lugo, Holes and Integrality of the Curvature, UCLA Preprint TEP/8, May 1976.

[35] W. Marciano and H. Pagels, Chiral Charge Conservation and Gauge Fields, Phys. Rev. (1976), idem and Z. Parsa, Multiply Charged Magnetic Monopoles, SU(3) Pseudoparticles and Gravitational Pseudoparticles, Rockefeller Preprint C00- 2232B-108, 1976.

[36] M. E. Mayer, Thesis, Unpublished, Univ. of Bucharest, 1956; Extended Invariance Properties of Quantized Fields, I, Preprint JINR, Dubna 1958 and Nuovo Cimento 11, 760 - 770 (1959).

[37] M. E. Mayer, C*-Bundles and Symmetries in Algebraic Quantum Field Theories, Proc. of Conf. on Noncompact Groups in Physics, Y. Chow, ed., Milwaukee, 1965; W. A. Benjamin, N. Y. 1966. Fibrations, Connections and Gauge Theories (An Afterthought to the Talk by A. Trautman, Proc. of the International Symposium on New Mathematical Methods in Physics, Bonn 1973, K. Bleuler and A. Reetz, eds. Talk at the Intern. Congr. of Mathematicians (Abstract N4), Vancouver, B. C., 1974. Gauge Fields as Quantized Connection Forms, Proc. of the Symposium on Differential-Geometrical Methods in Physics, Bonn, 1975, K. Bleuler and A. Reetz, eds., Lecture Notes in Mathematics, Springer Verlag, Berlin, 1977.

[38] M. E. Mayer, Gauge Fields and Characteristic Classes, to be published.

[39] A. A. Migdal, Rekursionnye uravneniya v kalibrovochnykh teoriyakh

polya (Recursion Ecuations in Gauge Field Theories), Zh. Eksp. Teor. Fiz. 69, 810 - 822 (1975) [Sov. Phys. JETP 42, 413-418 (1976)].

[40] E. Noether, Invariante Variationsprobleme, Nachr. Ges. Göttingen (math.-phys. Klasse) 1918, 235 - 257.

[41] V. I. Ogievetskii and I. V. Polubarinov, Zh. Eksp. Teor. Fiz. 41, 247 (1961) [Sov. Phys. JETP 14, 179 (1962)].

[42] K. Osterwalder, Gauge Theories on the Lattice, Cargese Lectures, 1976, to be published.

[43] A. M. Polyakov, JETP Lett. 20, 194 (1974); Sov. Phys. JETP 41, 988 (1975); Phys. Lett. 59B , 80, 82 (1975). Nordita Preprint, 1976.

[44] M. Prasad and C. Sommerfield, Phys. Rev. Lett. 35, 760 (1975).

[45] J. E. Roberts, Local Cohomology and Superselection, Preprint, 1976.

[46] A. Salam and J. C. Ward, Nuovo Cimento 11, 568 (1959).

[47] A. Salam, Nobel Symposium, 1968, Almquist & Wiksell, Stockholm.

[48] J. Schwinger, Phys. Rev. 82, 914 (1951); 125, 1043; 127, 324 (1962); 91, 714 (1953).

[49] I. E. Segal, Proc. Nat. Acad. Sci. USA 41, 1103 (1955); 42, 670 (1956). Quantization of Nonlinear Systems, J. Math. Phys. 1, 468 - 488 (1960); Quantized Differential Forms, Topology 7, 147 - 171 (1968).

[50] F. Strocchi, The Existence of Local Solutions to the Equations $\partial_\mu F^{\mu\nu} = j^\nu$ and $\Box\varphi = j$ in QFT, Princeton Seminar Notes 1971-72.

[51] F. Strocchi and A. S. Wightman, Proof of the Charge Superselection Rule in Local Relativistic Quantum Field Theory, J. Math. Phys. 15, 2198 - 2224 (1974).

[52] W. E. Thirring, Ann. Phys. (N. Y.) 16, 96 (1961).

[53] G. 't Hooft, Nucl. Phys. B33, 173; B35, 167 (1971).

[54] G. 't Hooft, Symmetry Breaking through Bell-Jackiw Anomalies, Phys. Rev. Lett. 37, 8 (1976). Computation of Quantum Effects due to a Four-Dimensional Pseudoparticle, Harvard Preprint, 1976. Cal-Tech Seminar, 1976.

[55] G. 't Hooft and M. Veltman, Nucl. Phys. B44, 189 (1973).

[56] A. Trautman, Infinitesimal Connections in Physics, Proc. Internat. Symposium on New Mathem. Methods in Physics, Bonn 1973, K. Bleuler and A. Reetz, eds., Bonn, 1973, and earlier work quoted there.

[57] R. Utiyama, Invariant Theoretical Interpretation of Interaction, Phys. Rev. 101, 1597 (1956).

[58] M. Veltman, Nucl. Phys. B21, 288 (1971).

[59] S. Weinberg, A Theory of Leptons, Phys. Rev. Lett. 19, 1264 (1967).

[60] S. Weinberg, Rev. Mod. Phys. 46, 255 (1974).

[61] H. Weyl, Gravitation und Elektrizität, Sber. Preuss. Akad. Wiss. 1918, 465 - 480; Z. Physik 56, 330 (1929).

[62] T. T. Wu and C. N. Yang, Concept of Nonintegrable Phase Factors and Global Formulation of Gauge Fields, Phys. Rev. 12D, 3845 (1975).

[63] C. N. Yang and R. L. Mills, Conservation of Isotopic Spin and Isotopic Gauge Invariance, Phys. Rev. 96, 191 (1954).

[64] C. N. Yang, Integral Formalism for Gauge Fields, Phys. Rev. Lett. 33, 445 - 447 (1974).

Cf. also the bibliography of the accompanying lecture notes by W. Drechsler.

ADDITIONAL REFERENCES

Albeverio, S. A., and Høegh-Krohn, R. J.: Mathematical Theory of Feynman Path Integrals, LNM 523, Springer Verl., 1976.

Feynman, R. P., and Hibbs, A. R.: Quantum Mechanics and Path Integrals, McGraw-Hill, New York, 1965

Palais, R. S.: Seminar on the Atiyah-Singer Index Theorem, Princeton Univ. Press, 1965.

[65] M. F. Atiyah, N. J. Hitchin, and I. M. Singer, Deformations of Instantons (Oxford-Berkeley-MIT Preprint, recd. May 1977).

[66] L. S. Brown, R. D. Carlitz, and C. Lee, Massless Excitations in Instanton Fields (U. of Washington Preprint, recd. May 1977).

[67] C. G. Callan,Jr., R. Dashen, and D. J. Gross, A Mechanism for Quark Confinement (IAS Preprint,C00-2220-94, 1977).

[68] V. De Alfaro, S. Fubini,and G. Furlan, A New Classical Solution for The Yang-Mills Equation, Phys. Lett. 65B, 163 (1976).

[69] S. Fubini, A New Approach to Conformal Invariant Field Theories, Nuovo Cimento 34A, 521 (1976). V. De Alfaro, S. Fubini,and G. Furlan, Conformal Invariance in Quantum Mechanics, Nuovo Cimento 34A, 569 (1976).

[70] R. Jackiw, C. Nohl, and C. Rebbi, Phys. Rev D (to appear), and R. Jackiw and C. Rebbi, Phys. Lett. B (to appear), both quoted in [65].

PART TWO

GAUGE THEORY OF STRONG AND ELECTROMAGNETIC INTERACTIONS FORMULATED ON A FIBER BUNDLE OF CARTAN TYPE

An Introduction to the Use of Differential Geometric Methods in Hadron Physics [x)]

x) Lectures presented in Mathematical Physics Lecture Series of the University of Texas at Austin, March 1977.

PART II

I. INTRODUCTION

It has always been of great interest in particle physics to study gauge theories as a means of introducing in a definite way new couplings of fields representing possible fundamental interactions between the elementary objects described by those fields. Following closely the well-known example of a gauge theory provided by spinor electrodynamics, i.e. electromagnetism in the presence of a quantum mechanical spinor matter wave function, which is characterized by the very simple abelian gauge group U(1), one usually proceeds by extending an invariance group G of a Lagrangian $\mathcal{L}^{(0)}$ (describing a certain supposedly known system of fields) to the invariance of a modified Lagrangian $\mathcal{L}'$ with respect to the gauge group G obtained by allowing the transformations of G to be space-time dependent. To insure G-gauge invariance the new Lagrangian $\mathcal{L}'$ must now contain, besides the fields entering $\mathcal{L}^{(0)}$, a set of additional fields, the so-called compensating fields - or gauge fields - ,which couple to the original fields in a well determined way introducing thereby an interaction between fields in an unambigous manner. It is probably fair to say that this gauge procedure[x)] of considering the transformation of a certain group G as locally different in different space-time points is not only a heuristically attractive method of introducing an additional interaction into a supposedly known dynamical system described by a Lagrangian $\mathcal{L}^{(0)}$, but that it is the method to establish interactions in particle physics in a non ad hoc way. From this point of view it would be highly satisfying if, indeed, all interactions in physics could be shown to be of gauge type. The weak interactions together with the electromagnetic interaction - the latter serving, as mentioned, as a model for the gauge interactions - is at present considered within the framework of a unified gauge theory of the Salam-Weinberg type[(1)] characterized by the gauge group $U(1) \otimes SU(2)$. The gravitational interaction, being described by a theory (general relativity) which is basically formulated in geometric terms, can also be given a gauge formulation[(2)]. Finally, for the strong forces the gauge concept has been an essential stimulus for research ever since its introduction by Yang and Mills[(3)] who extended the isospin symmetry of nuclear interactions to a SU(2) gauge theory for nuclear matter fields coupled to the Yang-Mills gauge fields.

It is now interesting to observe that any gauge theory can be characterized differential geometrically by using the concept of a fiber bundle. For physics mainly fiber bundles over a curved (in the presence of gravitation) or a flat hyperbolic space-time manifold are relevant. It is the choice of a certain bundle characterized by the choice of the structural group of this bundle which defines a geometric substratum characteristic of a certain interaction. Instead of basing the dynamics on a

---

x) The name gauge invariance, gauge fields etc. has a historical origin which we shall mention in Chapter II when we briefly refer to H. Weyl's unified theory of gravitational and electromagnetic interaction of 1918.

Lagrangian invariant with respect to a certain gauge group one could also start by choosing a certain geometry namely that fiber bundle geometry raised over space-time possessing a structural group identical with the gauge group of the Lagrangian formalism. Furthermore, one must say that the geometric framework in terms of fiber bundles is much more general. The conventional gauge trick based on a Lagrangian, which we shall discuss briefly in Chapter II by following the work of Utiyama[4], does not necessarily lead to all the physically interesting gauge theories since the Lagrangian from which one starts i.e. the Lagrangian $\mathcal{L}^{(0)}$ (without the gauge fields) which is invariant under transformations of a group G with constant parameters (global G-invariance) cannot be guessed so easily to exhaust all physically interesting possibilities for the associated gauge invariant theories. However, defining a certain gauge theory by choosing a fiber bundle with structural or gauge group G, motivated on physical as well as on geometrical grounds, one is always free to set up afterwards a Lagrangian formalism for such a theory which one could not have arrived at in the way discussed by Utiyama by allowing the parameters of a "known" symmetry group of a Lagrangian to become x-dependent. The further advantages of using a differential geometric language and differential geometric techniques for the description of the physics of gauge interactions are the following ones:

i) The fiber bundle formalism conveys an intuitive geometric meaning contained in notions like that of a connexion on the bundle space or that of a curvature etc. being, in fact, concepts which are more "anschaulich" than those appearing in a Lagrangian theory.

ii) The bundle formalism allows the discussion of global phenomena in gauge theories.

iii) The geometric or bundle formalism provides an ideal framework for a dualistic description of phenomena by coupling various quantities specifying the geometry to other quantities representing the matter distribution.

Since we want to give in this review a representation of strong interaction properties of matter in terms of differential geometric concepts and techniques we like to add some more remarks concerning point iii): The mutual interplay of matter and geometry, i.e. the idea that the distribution of matter affects - or stated more strongly - determines the geometry of the underlying space, was so convincingly demonstrated in general relativity establishing the geometric nature and origin of gravitation. Following this analogy we would like to ask the question whether this idea of matter influencing the underlying geometry could not be made to work once more again, however, this time not on a cosmological scale determining the geometry of the world in the large but on a microphysical scale determining the influence of a matter distribution on a suitably defined geometric substratum in the small i.e. in the immediate neighbourhood of this matter distribution. Our aim is to give strong interaction physics a basically geometric interpretation by choosing a certain fiber bundle constructed over space-time providing thereby not only a gauge description for hadronic interations but, furthermore, establishing a dualism of the mentioned kind between hadronic matter on the one hand and the underlying fiber bundle geometry on the other hand. The former is described by a generalized wave function or, more exactly, a generalized wave operator (see Chapter IV) being defined as a cross section on a fiber bundle possessing a structural or gauge group related to the dynamics of strong interaction, while the latter is described by connexion and curvature quantities on the bundle space. The aim is to account in this way for the extension of hadrons, i.e. the experimental fact that a proton or a neutron is not a point-like object treatable as a mass point endowed with spin moving according to the laws of relativistic quantum mechanics, but that hadrons are, in fact, extended structures possessing internal degrees of freedom which give rise to a whole mass and spin spectrum for these states. We picture here the internal motion of hadrons as being associated with degrees of freedom belonging to the fibers of the generalized space (the bundle space) used to characterize the underlying geometric substratum.

Now one could ask how it is possible to account for extension on a microphysical level without introducing an elementary length parameter into the description. The answer is that we do introduce an elementary length R of the order of 10^{-13} cm characterizing the range of the strong interaction forces. However, we do not introduce this length parameter into the geometry of space-time (the base space of our bundle) since we want to make use of continuum mathematics and apply differential geometric methods avoiding thus lattice gauge theories and the discussion of limits. Instead we shall characterize the fiber of the bundle over space-time by the length parameter R in choosing as fiber a Riemannian space of constant curvature with curvature radius R. Now immediately the question appears what measurable effects in space-time such a length parameter in the fiber could have since it is in the base space i.e. in space-time that one determines the extension of hadrons implied by the measured form factors for the proton and neutron. The answer here is that one has to choose a bundle of Cartan type over space-time, i.e. a bundle possessing a so-called Cartan connexion (see Chapter III). Such a bundle is characterized among other things by the fact that F_x, the local fiber over the space-time point x, is a space tangent to the base space at x. Hence, the coordinates of a point in the fiber — we shall call them ξ_x or simply ξ — can be regarded as generalized relative coordinates where "relative" means relative with respect to the point of contact of base space and fiber.

Before we discuss the theory in its mathematical details let us give in this introduction a rough qualitative picture of the dualism between hadronic matter and the underlying geometry referred to above which will be a central physical aspect of our discussion. The wave operator $\Psi(x,\xi)$, representing, say, a proton, will depend on space-time coordinates x and on the internal coordinates ξ varying in the local fiber over x. $\Psi(x,\xi)$ will possess representation properties with respect to the local Lorentz group in x [x)], and it will have representation properties with respect to the structural group G acting in the fiber. For a gauge description of strong interaction we shall choose G to be the SO(4,1) de Sitter group — or rather its covering group $\overline{SO(4,1)} = USp(2,2)$ since we want to consider operators possessing spinor representation character with respect to the internal de Sitter group. This ten parameter group acts as a group of motion in the fiber being a space which can be identified with the homogenous space SO(4,1)/SO(3,1). This noncompact coset space SO(4,1)/SO(3,1) is a Riemannian space, V_4', of constant (negative) curvature. All fiber bundles [xx)] of Cartan type over space-time with the standard fiber $F = V_4' \cong SO(4,1)/SO(3,1)$ and different radii of curvature of F are mathematically isomorphic, however, they are distinct structures from the point of view of physics. As mentioned, we choose, in view of strong interactions, a curvature radius R of F of the order of 10^{-13} cm characterizing thus the fiber of the bundle (and thereby the bundle space as a whole) by this length parameter in a similar way as the base space of the bundle is characterized by the velocity of light, $c=3 \cdot 10^{10}$ cm/sec, specifying the local Lorentz structure in each Minkowski tangent space to a general curved space-time.

Now $\Psi(x,\xi)$ — given as a cross section on a vector bundle associated with a prinicipal frame bundle over space-time having the structural group SO(4,1) (see Chapter III) — can be factorized into a conventional q-number Dirac space-time part $\varphi(x)$ (for spin 1/2 baryons), and an internal, i.e. de Sitter part $\phi_x(\xi)$. Here $\phi_x(\xi)$ transforms under the gauge group G = SO(4,1) (or $G = \overline{SO(4,1)} = USp(2,2)$ for a spinor) and couples to the connexion and curvature fields defined on the bundle space specifying in detail the geometry in what we called the substratum [xxx)]. In analogy to the situation in electromagnetism where a certain current distribution of

---

x) Or the global Lorentz group in case the base space is flat Minkowski space-time.

xx) For the details of the definition of bundles with Cartan connexion see Chapter III.

xxx) We like to remark in passing that in order to represent also more local objects, i.e. leptons, in a world possessing basically the geometry of a fiber bundle over space-time one has to represent them as internal scalars or as a constant (trivial representation). Fields with such a representation character with respect to the internal or gauge group G = SO(4,1) do not couple to the bundle connexion or the associated internal curvature and, hence, would not experience any effects mediated through the fibers. In the geometric description of strong interactions we are aiming at we would then say that such objects would not "feel" the fiber structure i.e. do possess neither strong interaction nor extension (the latter at least not on the level of the parameter R).

charged matter sets up the electromagnetic fields; as well as in analogy to general relativity where a certain energy and momentum distribution of matter sets up the metric and curvature fields, we shall now suppose that there is a certain hadronic current operator, bilinear in the fields $\psi(x,\xi)$ and its adjoint (to be defined in Chapter IV), acting as a source for the curvature and connexion fields on the Cartan-de Sitter bundle space. We call the equations connecting geometric and matter quantities the current-curvature equations and refer to the dynamics implied by this gauge description for the strong interactions as to the strong fiber dynamics (SFD). In this scheme the geometrical quantities, specifying the local geometry in detail, are regarded as induced locally by the matter distribution resulting thus locally in a deviation from a flat space possessing zero internal curvature [x)].

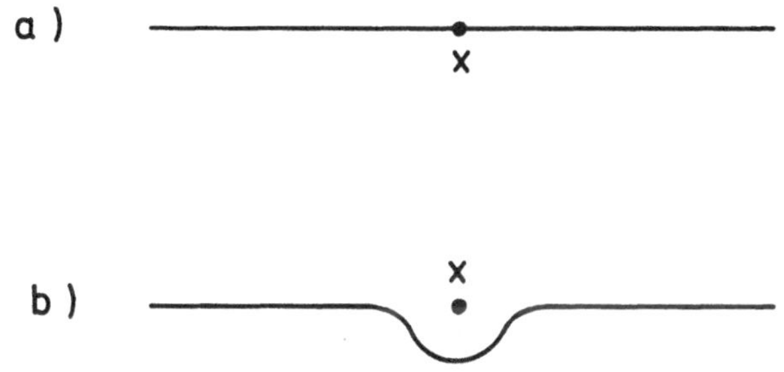

a) point - like description
b) fiber bundle description

Fig. 1

The situation is schematically drawn in Fig. 1. Case a) represents the conventional description in terms of a local operator quantum field defined on an inert background geometry providing merely the arena for the point-like physics to be considered. Case b) shows in a very schematic way the fiber bundle description (SFD) where the point spinor part (symbolized by a dot) is accompanied by gauge fields induced locally in the geometry in a smooth fashion distorting the geometry locally by setting up a nonzero internal curvature preventing thus the substratum on which hadron physics is thought to evolve in this description from being a passive arena representable by a flat background space. Here the response in the geometry is an essential part of the phenomenon making up a hadron. In fact, the extension of hadrons appears here as the combined effect of a local quantum field $\psi(x)$ dressed, so to

---

[x)] By "internal" curvature we mean the curvature associated with the Cartan-de Sitter fiber structure.

speak, with a halo of gauge fields induced in the underlying fiber bundle geometry which is related to the factor $\phi_x(\xi)$ of $\psi(x,\xi)$.

At this level of the description the connexion and curvature fields on the bundle space are considered as classical c-number gauge fields. Only the point spinor [x)] part of $\psi(x,\xi)$ appears as a second quantized operator field. This scope will have to be broadened when solutions refering to certain matrix elements of the current-curvature equations are known. It is to be expected that also the gauge fields will develop a q-number component due to the fact that these fields are coupled to a second quantized source current. However, for the purpose of the discussion in these lectures we assume an intermediate position and regard the fields induced in the geometric stratum as purely classical fields. This is the domain where the bundle formalism provides an intuitive guide. The quantum problem for the gauge fields goes beyond this geometric picture and must be solved using additional means. Thus, for the discussion given here the internal part $\phi_x(\xi)$ of $\psi(x,\xi)$, representing the "motion in the fiber over x", is a smooth function of ξ and x, since a distribution in ξ cannot disappear discontinuously in x on the bundle space. This gauge part of $\psi(x,\xi)$ is thus expected to give rise to formfactor effects.

The proposed dualism of hadronic matter influencing the underlying fiber bundle geometry with material particles leaving an imprint on this geometry provides a gauge description for hadrons and their interaction avoiding the constituent puzzle of present day hadron models. This imprint, or halo as we called it before, is the integral effect of the local gauge fields set up in the Cartan bundle geometry characterized as a whole by the length parameter R. The halo in the geometry accompanying a baryonic particle — or, more exactly, being part of the description of a hadron — cannot be separated into constituent parts, at least not on a classical level. However, the question of a granular structure of these gauge field contributions making up an extended hadron leads us back to the question of the quanta for these gauge fields which is disregarded here and, in fact, not yet fully understood. To the question whether there are internal modes for the "motion in the fiber" which act under certain kinematical conditions as some kind of partons simulating a more local structure in hadronic interactions than implied by our bundle formalism with an elementary length parameter of the order of one Fermi built into the geometry we have no simple answer at present. This part of our description of hadronic phenomena has still to be explored. Let us, therefore, postpone the difficult quantum problem for the gauge fields to future investigations and focus here the attention on the problem of setting up a reasonable geometric description for extended hadrons immersed in a geometry having basically the structure of a fiber bundle over space-time with a Cartan connexion.

---

x) For a spin 1/2 baryon.

II. GAUGE THEORIES IN A LAGRANGIAN FORMULATION

In this chapter we briefly review the traditional approach to gauge theories bases on a Lagrangian density expressed in terms of local fields and their first derivatives. We closely follow in section II.2 the presentation given by Utiyama[(4)] modelled after the well-known formulation for the interaction of electromagnetic fields with matter (i.e. electrons) represented by a quantum mechanical wave function and the extension of the gauge concept to strong interactions as suggested by the work of Yang and Mills[(3)]. For compelteness and as a transparent example we give in the first section of this chapter a short discussion of electromagnetism as a gauge theory and translate the notions appearing there into a geometric language which is suggestive in connection with a possible generalization of the gauge concept to other less understood interactions in physics.

II.1 Spinor Electrodynamics

In quantum mechanics the wave function $\psi(x)$ describing the motion of an electron can be subjected to the following phase transformation

$$\psi(x) \longrightarrow \psi'(x) = e^{-i\alpha}\,\psi(x) \qquad (2.1)$$

where $\psi(x)$ denotes in the nonrelativistic case the Schrödinger wave function and in the relativistic case the four component Dirac wave function, and α represents an arbitrary constant phase angle. The expectation value of an operator representing a measurable quantity,or the bilinear probability $W(x)\,d^3x = \psi^*(x)\,\psi(x)\,d^3x$ derived from $\psi(x)$, i.e. the probability of finding the electron in a volume element d^3x at the point x, are unchanged by the transformation (2.1). $\psi(x)$ and $\psi'(x)$ are said to describe the same physical state and one speaks of a ray representation in the Hilbert space of states.

An even more far reaching freedom in choosing field quantities characterizing the measurable physical fields is realized in classical electromagnetism where the electric and magnetic fields $\vec{E} = (F_{10}, F_{20}, F_{30})$ and $\vec{H} = (F_{23}, F_{31}, F_{12})$ can be derived from a potential $A^\mu(x) = (A^0(x) = \phi(x), \vec{A}(x))$ in the well-known way x)

---

x) $A^0(x) = \phi(x)$ is the scalar potential, $\vec{A}(x)$ the vector potential. We use here the notation of Bjorken and Drell[(5)] with $x^\mu = (x^0, \vec{x})$, $x_\mu = \eta_{\mu\nu}x^\nu$ and, analogously, $A_\mu(x) = \eta_{\mu\nu}A^\nu(x)$ where $\eta_{\mu\nu}$= diag (1,-1,-1,-1). $\partial_\mu = \partial/\partial x^\mu = (\partial_0, \vec{\nabla})$. The momentum operator in quantum mechanics is given in the Schrödinger representation by $\hat{p}_\mu = i\partial_\mu$. We shall use units in which $\hbar$ = c = 1 except for a few occasions which are explicitely mentioned.

$$F_{\mu\nu}(x) = \partial_\nu A_\mu(x) - \partial_\mu A_\nu(x) \tag{2.2}$$

with the potentials being merely computational aids which can be changed by a so-called gauge transformation of the first kind,

$$A_\mu(x) \longrightarrow A'_\mu(x) = A_\mu(x) + \partial_\mu \lambda(x) \tag{2.3}$$

without altering the electromagnetic field strengths $F_{\mu\nu}(x)$. To satisfy the Lorentz condition, $\partial^\mu A_\mu(x) = 0$, before and after a transformation of the type (2.3) the function $\lambda(x)$ is arbitrary except for the requirement that it has to satisfy the equation $-\Box\lambda(x) = \partial^\mu \partial_\mu \lambda(x) = 0$.

Let us now focus attention on the description of electromagnetic fields in the presence of charged matter described by a quantum mechanical wave function $\psi(x)$. To be definite we shall regard $\psi(x)$ to be a four component Dirac spinor wave function describing electrons. It is well known that in the quantum mechanical description the potentials $A_\mu(x)$ do appear in the fundamental equations of motion (the Dirac equation) showing that the fields $A_\mu(x)$ are of physical relevance and not merely computational aids as they were on the classical level. Now the question arises what is actually a complete description of electromagnetic effects in the presence of matter in the form of electrons described quantum mechanically in terms of a wave function $\psi(x)$. Are the A_μ-fields, indeed, measurable quantities or are only the $F_{\mu\nu}$ of classical electrodynamics determinable; is the phase of the wave function measurable or only the complex phase factor [x)xx)]

$$e^{-\frac{ie}{\hbar c}\int_x^y A_\mu(x')\,dx'^\mu} \tag{2.4}$$

taken along a definite path joining the points x and y as suggested by Aharonov and Bohm[(8)]? We shall come back to the question what constitutes a complete description of electromagnetic effects in spinor electrodynamics after we have briefly recapitalated the usual arguments.

The free Dirac equation follows as Euler-Lagrange equation from the variational principle

$$\delta \int_\Omega \mathcal{L}^{(0)}(x)\, d^4x = 0 \tag{2.5}$$

---

x) Compare in this context the work of Mandelstam[(6)] and Wu and Yang[(7)] to which we return below.

xx) For clarity we have written here the phase in the conventional way by including the factors $\hbar$ and c. e is the electric charge.

with Ω denoting a space-time domain, and with the Lagrangian $\mathcal{L}^{(0)}(x)$ given by

$$\mathcal{L}^{(0)}(x) = \bar{\psi}(x)\, i\gamma^{\mu}\partial_{\mu}\psi(x) - m\,\bar{\psi}(x)\psi(x)\,. \tag{2.6}$$

Here γ^{μ} are the four Dirac matrices obeying

$$\{\gamma^{\mu},\gamma^{\nu}\} = 2\eta^{\mu\nu}\cdot 1 \quad ; \quad \gamma^{\mu\dagger} = \gamma^{0}\gamma^{\mu}\gamma^{0}\,, \tag{2.7}$$

m is the mass of the particle described by $\psi(x)$ and $\bar{\psi}(x) = \psi^{\dagger}(x)\gamma^{0}$. Eq. (2.6) is seen to be invariant under the U(1)-phase transformations (2.1).

In classical mechanics the canonical generalized momentum and energy variables used to describe the motion of a single charged mass point in the presence of electromagnetic fields involve the potentials $A^{\mu}(x)$ according to x)

$$E - e\,\phi(x) \tag{2.8a}$$

$$\vec{p} - e\vec{A}(x) \tag{2.8b}$$

which, in the transition to quantum mechanics, leads in eq. (2.6) to the well-known replacement

$$\partial_{\mu} \longrightarrow \partial_{\mu} + ie\,A_{\mu}(x) \tag{2.9}$$

called the minimal electromagnetic interaction. Carying out this minimal replacement in $\mathcal{L}^{(0)}(x)$ and adding the Lagrangian density for the free electromagnetic fields,

$$\mathcal{L}^{e.m.}(x) = -\tfrac{1}{4}F_{\mu\nu}(x)\,F^{\mu\nu}(x) \tag{2.10}$$

one obtains the Lagrangian

$$\begin{aligned}\mathcal{L}(x) = \mathcal{L}'(x) + \mathcal{L}^{e.m}(x) &= i\,\bar{\psi}(x)\gamma^{\mu}\left[\partial_{\mu} + ieA_{\mu}(x)\right]\psi(x) - \\ &- m\,\bar{\psi}(x)\psi(x) - \tfrac{1}{4}F_{\mu\nu}(x)F^{\mu\nu}(x)\end{aligned} \tag{2.11}$$

---

x) $e = -|e|$ for an electron; $p^{\mu} = (p^{0} = E, \vec{p})$ being the relativistic four-momentum.

of spinor electrodynamics providing a correct description of the electromagnetic effects at the level of atomic physics and representing — in its second quantized form — the basis of quantum electrodynamics (QED) [x)]. In the well-known manner one deduces from the variational principle $\delta\int_\Omega \mathcal{L}(x)d^4x = 0$, with $\mathcal{L}(x)$ as given by eq. (2.11), the Dirac-Maxwell equations of motion

$$\gamma^\mu \left[\partial_\mu + ieA_\mu(x)\right]\psi(x) = -im\,\psi(x) \tag{2.12}$$

$$\partial^\nu F_{\mu\nu}(x) = j_\mu(x) \tag{2.13}$$

with the source current $j_\mu(x)$ being given by

$$-\frac{\partial\mathcal{L}(x)}{\partial A^\mu(x)} = j_\mu(x) = e\,\bar{\psi}(x)\gamma_\mu\psi(x) \tag{2.14}$$

which, from eqs. (2.12) and its adjoint or directly from eq. (2.13) (due to the antisymmetry of the $F_{\mu\nu}$), is seen to be conserved, i.e.

$$\partial^\mu j_\mu(x) = 0 \tag{2.15}$$

expressing the conservation for the electric charge.

The second group of Maxwell equations, i.e.

$$\partial_{\{\varkappa} F_{\mu\nu\}}(x)_{\text{cycl.}} = \partial_\varkappa F_{\mu\nu}(x) + \partial_\mu F_{\nu\varkappa}(x) + \partial_\nu F_{\varkappa\mu}(x) = 0 \tag{2.16}$$

follows directly from the definition (2.2) of the electromagnetic fields in terms of the potentials and represent integrability conditions which in the geometric language adopted later correspond to the Bianchi identities for the curvature tensor (see below for the interpretation of the $F_{\mu\nu}(x)$ as a curvature tensor).

Characterizing now the Lagrangian (2.11) by an invariance one early noticed in the development of relativistic quantum mechanics that $\mathcal{L}(x)$ is invariant under the combined transformations (compare eqs. (2.1) and (2.3)) [xx)]

---

x) We do not consider the quantum field aspect in this section. For a second quantized treatment see, for example, ref. 5.

xx) For reference to the older literature we quote H. Weyl[(9)] and F. London[(10)]. To Weyl's paper of 1918 we shall return in more detail at the end of this section.

$$\psi(x) \longrightarrow \psi'(x) = e^{-ie\alpha(x)}\psi(x) \tag{2.17a}$$

$$A_\mu(x) \longrightarrow A'_\mu(x) = A_\mu(x) + \partial_\mu \alpha(x) \tag{2.17b}$$

with $\alpha(x)$ being now an arbitrary real space and time dependent function restricted only by requiring that $\Box\alpha(x)=0$. The transformations (2.17) are referred to in the literature as the gauge transformations of the second kind, and the invariance of eqs. (2.11) - (2.16) under (2.17) as the gauge invariance of the theory.

Having found an invariance property of a dynamical system composed of the fields $\psi(x)$ and $A_\mu(x)$ which is known to represent nature one could now turn the argument around and ask the following question: Given the U(1) phase-invariant Lagrangian $\mathcal{L}^{(0)}(x)$ of eq. (2.6). Extending the transformations (2.1) to x-dependent U(1) transformations - now called U(1) gauge transformations - and demanding invariance of the Lagrangian under these transformations one can ask what kind of new fields have to be introduced into the theory described by $\mathcal{L}^{(0)}(x)$ such that the extended invariance postulate is indeed satisfied. The well-known answer here is that one has to introduce just the four fields $A_\mu(x)$, called the gauge potentials, transforming inhomogeneously as shown in eq. (2.17b), in order to compensate the term originating from the differentiation of the factor $e^{-ie\alpha(x)}$ — hence the name compensating fields for these gauge potentials. Additional phenomena connected with the new fields can in principle also appear disconnected from the ψ-field, i.e. the gauge fields give rise to an energy density of their own contributing thus to the total Lagrangian a term which is given in a gauge invariant way by eq. (2.10) which is the simplest expression written down in terms of the gauge invariant field strengths (2.2) following from the potentials $A_\mu(x)$.

By an analogous line of reasoning one has now a method at hand of introducing a gauge interaction into a system of fields described dynamically by a certain Lagrangian $\mathcal{L}^{(0)}(x)$ possessing an invariance group G referring to a global G-invariance of the system. Postulating now a local G-invariance by allowing the parameters determining the transformations of the group G to become x-dependent functions, i.e. demanding G-gauge invariance of the theory, results in the extension of the principle of minimal electromagnetic interaction with its definite form of the coupling between the quantum mechanical ψ-function and the electromagnetic potentials to a new type of interaction in physics characterized by another group G different from the group U(1) associated with electromagnetism.

Before we describe this gauge trick for an arbitrary group G in the framework of a Lagrangian formulation which leads to an interaction Lagrangian analogous to the form $-j_\mu(x)A^\mu(x) = -e\,\bar{\psi}(x)\gamma_\mu\psi(x)A^\mu(x)$ characteristic of spinor electrodynamics (compare eqs. (2.11) and (2.14)) let us, however, first insert here a few remarks concerning the geometric interpretation of the U(1) gauge invariance of the electromagnetic interaction. Although we defer the exact definition of the differential geometric concepts involved to the next chapter we like to reformulate in geometric terms here the notions appearing in the U(1) gauge theory described above and mention some of its peculiar properties. This theory implies that the wave function $\psi(x)$ representing matter — or, more exactly, point-like electrons — is actually not an ordinary function defined on space-time but more properly described as a cross section on a fiber bundle constructed over space-time possessing the fiber and structural group U(1). The relevant fiber bundle is given locally by a direct product of space-time and the unit circle. By a cross section on this bundle is ment the choice of a phase angle on the unit circle for the wave function at each space-time point x in a smooth fashion for all x [x)]. The U(1) gauge invariant derivative

$$D_\mu = \partial_\mu + ie\,A_\mu(x) \tag{2.18}$$

obtained after performing the "minimal replacement" (2.9) is just the operator for the covariant derivative on the U(1) bundle with the electromagnetic potentials $A_\mu(x)$ playing the role of the coefficients of a connexion on the bundle. The $A_\mu(x)$ possess the inhomogeneous transformation character (2.17b) associated with a change of the gauge expressed by (2.17a) corresponding to a transition to another cross section for the wave function $\psi(x)$ related to the previous one by changing the local phase according to eq. (2.17a). The electromagnetic field strengths $F_{\mu\nu}(x)$, being gauge independent quantities defined by eq. (2.2), represent the components of the U(1) gauge curvature tensor associated with the connexion defined by the $A_\mu(x)$ being constrained by the Bianchi identities (2.16). The fact that the generalized space on which the wave function is defined (i.e. the U(1) bundle space over space-time) possesses a curvature is immediately seen from the noncommutativity of two successive covariant derivatives applied to $\psi(x)$ i.e.

$$[D_\mu, D_\nu]\,\psi(x) = -ie\,F_{\mu\nu}(x)\,\psi(x). \tag{2.19}$$

[x)] A more proper mathematical definition of a fiber bundle and a cross section on it as well as the definition of a connexion on a fiber bundle will be given in the next chapter. Moreover, electromagnetism, without magnetic monopoles, is characterized by a so-called trivial U(1) bundle which is globally the direct product of space-time and the unit circle.

Following Mandelstam[6] one can now go over from the gauge dependent fields $\psi(x)$ and $\bar{\psi}(x)$ to new fields $\phi(x,P)$ and $\bar{\phi}(x,P)$ yielding a gauge independent but path dependent description of electromagnetic phenomena and electron fields. These path dependent fields are defined by

$$\phi(x,P) = \psi(x)\, e^{ie\int_{-\infty}^{x} A_\mu(x')dx'^\mu} \tag{2.20a}$$

$$\bar{\phi}(x,P) = \bar{\psi}(x)\, e^{-ie\int_{-\infty}^{x} A_\mu(x')dx'^\mu} \tag{2.20b}$$

where the line integral appearing in the expontentials are taken over a space-like path P in Minkowski space running from minus infinity to the point x. The right-hand side of eqs. (2.20) are clearly gauge independent as one immediately sees by making the replacements (2.17) and integrating the complete differential originating from the gradient term using the boundary condition $\alpha(-\infty)=0$. The exponential factors in eqs. (2.20), representing so to speak the integral over the gauge field histories taken along a definite path P , is analogous to the phase factor (2.4) mentioned previously being an element of the group U(1) associated with the path P. The condition that the connexion on the U(1) bundle is integrable, i.e. that the phase factor is path independent, corresponds to the vanishing of the tensor $F_{\mu\nu}(x)$ implying that the U(1) bundle space admits a flat connexion. In general there is a non-zero U(1) gauge curvature present (electromagnetic fields) so that (2.4) represents a nonintegrable phase factor to use the terminology of Wu and Yang[7]. It is easy to show that the equation

$$\partial_\mu \phi(x,P) = 0 \tag{2.21}$$

expressing the constancy of the path dependent field $\phi(x,P)$, is equivalent to the equation

$$D_\mu \psi(x) = 0 \tag{2.22}$$

for $\psi(x)$ which can be interpreted geometrically as the equations for the parallel shift of the Dirac wave function $\psi(x)$ with respect to the connexion determined by $A_\mu(x)$.

Let us finally, for a discussion of the Aharonov-Bohm experiment given below, take an intermediate position between Mandelstam's point of view of using a path dependent electron field $\phi(x,P)$ and the conventional view of using a gauge dependent electron field $\psi(x)$. Starting from the assumption that we are able to prepare an electron beam with a certain phase at a point x we ask the question what the phase of that beam would be if the electrons in the beam were brought to a different point y a finite distance away along a path C_{yx} joining x and y (directed from x to y) in a region of space-time where $A_\mu(x')$ is non-zero. Let us denote the wave function at the point y obtained by integraging eq. (2.22) along the curve C_{yx} (with the initial value $\psi(x)$ for y = x) by $\psi(y, C_{yx})$. Then one has

$$\psi(y, C_{yx}) = S(C_{yx})\,\psi(x) \tag{2.23}$$

where

$$S(C_{yx}) = e^{-ie\int_y^x A_\mu(x')dx'^\mu} \tag{2.24}$$

being the nonintegrable phase factor associated with the path C_{yx} . $S(C_{yx})$ is identical to the U(1) transformation in the fiber bundle over space-time characterizing the electromagnetic interaction which is associated with the unique path in the bundle space, called the horizontal lift of the curve C_{yx}, defining thereby the parallel transport of $\psi(x)$ along a lifted curve on the U(1) bundle with respect to a given connexion. Physically the effect of the phase factor $S(C_{yx})$ can be measured by the Aharonov-Bohm experiment[8] in letting an electron beam go around a small cylindrical region D (see Fig. 2) containing a confined magnetic flux and observe the

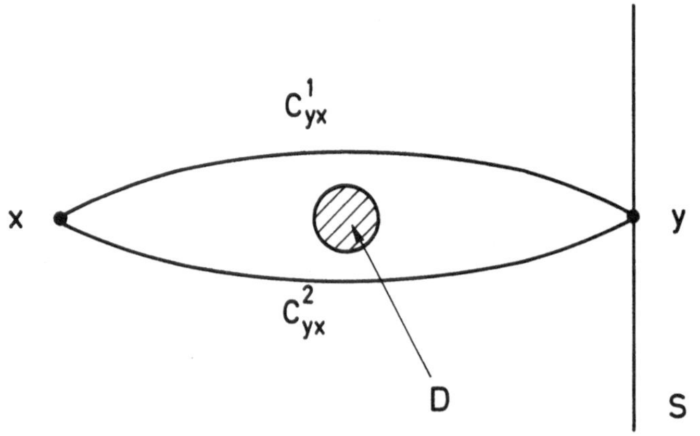

x : Electron source

S : Screen

D : Cylindrical region with { a) $F_{\mu\nu} \neq 0$ b) $F_{\mu\nu} = 0$

Fig. 2

change of the interference pattern noticible on the screen for a) $F_{\mu\nu} \neq 0$, and b) $F_{\mu\nu} = 0$ in the domain D. At the location of the actual paths C^1_{yx} and C^2_{yx} it is experimentally secured that there are no fields $F_{\mu\nu}(x)$ present which could interact directly (via the Lorentz force law) with the electrons in the beams. Thus the electrons "see" only the potentials along their paths and these potentials are different if they correspond to a nonintegrable connexion (nonzero curvature, case a)) or if they correspond to an integrable connexion (zero curvature, case b)). It has been established experimentally by Chambers[11] that it is exactly the information contained in the phase factor (2.24) which is measurable experimentally, i.e. the difference in phase of the electron wave functions associated with the paths C^1_{yx} and C^2_{yx} determining the interference pattern at y on the screen is given by [x)]

$$S(C^1_{yx})S(C^2_{xy}) = S(C_x) = e^{-ie\frac{1}{2}\int_F F_{\mu\nu}(x')d\sigma^{\mu\nu}} \tag{2.25}$$

where we have denoted by C_x the closed contour $C^1_{yx} + C^2_{xy} = C^1_{yx} - C^2_{yx}$ and by F an arbitrary surface containing the contour C_x with $d\sigma^{\mu\nu}$ being the surface element on F. Thus the question which quantities provide a complete description of electromagnetic effects in the presence of charged matter (electrons) described by a quantum mechanical wave function $\psi(x)$ is to be answered in the following way: The $F_{\mu\nu}(x)$ entering the Lorentz force law <u>and</u> the phase factor $S(C_{yx})$ governing the interference behaviour of matter in a quantum mechanical treatment provide a full description of the measurable electromagnetic effects associated with the physics of electrons in quantum theory.

Let us finally mention that if one performs a gauge transformation (2.17) in eq. (2.23), i.e. represents the electron beam in the Aharonov-Bohm experiment by another cross section (a gauge transformed ψ-function) one has

$$\psi'(y, C_{yx}) = S'(C_{yx})\,\psi'(x) \tag{2.26}$$

---

x) For the Aharonov-Bohm experiment described, with the region D containing a magnetic flux, the quantity (2.25) reduces to $\exp(-\frac{ie}{\hbar c}\oint \vec{A}\,d\vec{s}) = \exp(-\frac{ie}{\hbar c}\int_F \vec{H}\,d\vec{f})$ (where we have re-inserted the constants $\hbar$ and c). As an addition we remark that from this last mentioned formula it is only a small step to deduce Dirac's quantization condition for the electric charge[12] in case of existence of a single magnetic monopole of pole strength g in nature. Making use of the arbitrariness of the surface F containing C_x and using $\int_{F-F'} \vec{H}\,d\vec{f} = \int_V \operatorname{div}\vec{H}\,d\tau = 4\pi g$ [with V being the volume between F and another surface F'] the continuity of the phase factor, i.e. the continuity of the quantum mechanical wave function, in pushing the surface F across the monopole into the position F' requires that $4\pi i\, eg/(\hbar c) = \pm 2\pi i n$ with n being a positive integer. This corresponds to the quantization condition $e = \frac{\hbar c}{g}\cdot\frac{n}{2}$; n = ± 1, ± 2, ± 3 ... for the electric charge.

with
$$S'(C_{yx}) = S(y)\, S(C_{yx})\, S^{-1}(x) \tag{2.27}$$

where
$$S(x) = e^{-ie\alpha(x)} . \tag{2.28}$$

Eq. (2.17b) can be written in terms of $S(x)$ as

$$A'_{\mu}(x) = A_{\mu}(x) - \frac{i}{e}\, S(x)\, \partial_{\mu} S^{-1}(x) . \tag{2.29}$$

Eq. (2.27) shows that the phase factor suffers a gauge transformation at the endpoints of the path C_{yx} which is in accord with the group property obeyed by the $S(C_{yx})$ i.e.

$$S(C_{zx}) = S(C_{zy})\, S(C_{yx}) \tag{2.30}$$

where y is a point on the curve C_{zx} joining x and z. Furthermore, one has in an obvious notation

$$S(C_{xy}) = [S(C_{yx})]^{-1} \tag{2.31}$$

and
$$S(C_{xx}) = 1 . \tag{2.32}$$

In closing this section we briefly mention Hermann Weyl's[9] original form of gauge invariance of 1918 which originated from the idea of a nonintegrable i.e. path-dependent transfer of the standard of length in the curved space-time manifold of general relativity.

The transfer of a vector along a path in a curved Riemannian space is a nonintegrable process i.e. depends on the path used. Taking a vector around a closed loop C_x starting from the point x and returning to this point the vector displaced along C_x will be rotated compared to its original direction at the point x. For an infinitesimal closed loop being the borderline of an infinitesimally small portion of the space (which is a measurable quantity, i.e. possesses a certain area for a Riemannian space) the rotation of the vector per unit area is a measure of the local curvature of the space. In general relativity the underlying space is the curved hyperbolic Riemannian space-time V_4, and it was the ingenious idea of Einstein to interprete the local curvature field as a field describing gravitational effects arising from distant sources (mass distributions). In a Riemannian space the connexion coefficients and the curvature tensor (see chapter III) can be expressed in terms of the metric tensor $g_{\mu\nu}(x)$ alone, the latter playing thus physically the role of gravitational potentials defining geometrically a line element through the bilinear form

$$ds^2 = g_{\mu\nu}(x)\, dx^\mu \otimes dx^\nu. \tag{2.33}$$

In extending now the idea of giving the long range gravitational field a geometric interpretation Weyl suggested to incorporate electromagnetism, the other known long range field, together with gravitation into an enlarged geometric scheme such that both these long range fields in nature have a common geometric origin. Weyl proposed that the transfer of the unit of length relative to which the length of a vector transferred from one point of the manifold to another is measured should also be a nonintegrable process. In taking a vector around a closed loop C_x this would mean that upon arrival at x, after transfer along C_x, the vector is not only rotated but also to be referred to a different standard of length. In order to describe this mathematically Weyl introduced, in addition to the metric fundamental form (2.33), the linear form

$$d\ell = \ell\, A_\mu(x)\, dx^\mu \tag{2.34}$$

implying that the change, $d\ell$, of the unit of length, ℓ, is proportional to ℓ and proportional to the local coordinate differentials dx^μ multiplied by a set of functions $A_\mu(x)$. Weyl showed that these functions possess all the properties of the electromagnetic potentials. Instead of combining the gauge transformation of the $A_\mu(x)$ with the (in 1918 still unknown) quantum mechanical wave function $\psi(x)$ according to eqs. (2.17), multiplying thus $\psi(x)$ locally by a complex factor of absolute value one, Weyl suggested to combine the gauge transformation of the $A_\mu(x)$ with a transformation of the metric $g_{\mu\nu}(x)$ in space-time giving it a local real factor according to the replacements:

$$g_{\mu\nu}(x) \longrightarrow g'_{\mu\nu}(x) = e^{\lambda(x)} g_{\mu\nu}(x) \tag{2.35}$$

$$A_\mu(x) \longrightarrow A'_\mu(x) = A_\mu(x) - \partial_\mu \lambda(x) \tag{2.36}$$

These equations are Weyl's original gauge transformations of 1918 which really imply a local change of scale, i.e. a gauging of the standard of length, which explains the name he gave to these transformations. Integrating eq. (2.34) along a curve joining x and y one obtains

$$\ell_y = \ell_x \, e^{\int_x^y A_\mu(x')dx'^\mu} \tag{2.37}$$

where ℓ_y and ℓ_x are the units of length at the points y and x, respectively, and the exponential represents the nonintegrable real factor associated with the path C_{yx} used to translate the corresponding units of length.

Despite its inherent beauty Weyl's pre-quantum mechanical gauge theory trying to unify gravitational and electromagnetic interactions in an enlarged geometric framework going beyond Riemannian geometry was in this form untenable. A. Einstein to whom Weyl had submitted his paper for presentation to the Prussian Academy of Science wrote a short addendum to it in which he pointed out that the existence of sharp spectral lines in the spectrum of atoms having had different histories, i.e. having passed through different electromagnetic potentials, makes it extremely unlikely that such a nonintegrability of the unit of length is realized in nature. It was, to my knowledge, F. London[(10)] who first — after quantum mechanics was established — reconsidered Weyl's theory and turned the concept of gauge invariance into the form (2.17) under which we know it today.

II.2 G-Gauge Invariant Lagrangian Formalism

Extending the gauge description for the electromagnetic interaction as reviewed in the previous section to an arbitrary n-parameter gauge group G characterizing possibly other interactions in nature one starts, in analogy to what has been said in Section I.1, from a Lagrangian $\mathcal{L}^{(0)}(x)$ assumed to be known and invariant under transformations of a group G with constant parameters α_a ; $a = 1,2, \dots$ n (global G-invariance). Then one extends the G-symmetry of $\mathcal{L}^{(0)}(x)$ to the G-gauge invariance of a new Lagrangian $\mathcal{L}'(x)$, containing additional fields, by demanding invariance against x-dependent transformations of G determined by space and time dependent group parameters

$$G: \quad \alpha_1(x)\,,\, \alpha_2(x),\, \dots \quad \alpha_n(x). \tag{2.38}$$

To secure gauge invariance a set of 4•n new fields, the G-gauge potentials,have then to be introduced which enter the operator for a G-gauge invariant derivative representing the "minimal coupling" which characterizes the gauge theory based on the group G. An essential point here is to remark that one has to know a certain dynamical system described by a Lagrangian $\mathcal{L}^{(0)}(x)$ together with its invariance group G to start with. If this is the case then the method described by Utiyama[(4)] and reviewed in this section leads to a theory possessing an additional gauge interaction associated with the group G. We shall see later that not every physically interesting gauge theory can be obtained in this way simply because the starting point - the Lagrangian $\mathcal{L}^{(0)}(x)$ - cannot be guessed so easily. Basing, on the other side, the arguments from the beginning on a geometric ground by using a fiber bundle formalism with structural group G, as will be explained in Chapter IV for the strong interactions, then one at once arrives at a gauge theory without having to know the dynamics described by a Lagrangian $\mathcal{L}^{(0)}(x)$ beforehand [x)]. Let us, however, for the discussion in this section assume that the Lagrangian $\mathcal{L}^{(0)}(x)$ were, indeed, known and G-invariant describing a physically interesting theory.

Due to the assumed global G-invariance $\mathcal{L}^{(0)}(x)$ is composed of a set of fields $Q^A(x)$; $A = 1,2, \ldots N$ (with x denoting a point in flat Minkowski space-time) and their first derivatives, $\partial_\mu Q^A(x) = Q^A_{,\mu}(x)$, which transform as representations of the group G. Formulating this for infinitesimal transformations of G one has [xx)]

$$Q^A(x) \longrightarrow Q'^A(x) = Q^A(x) + \delta Q^A(x) \tag{2.39a}$$

$$Q^A_{,\mu}(x) \longrightarrow Q'^A_{,\mu}(x) = Q^A_{,\mu}(x) + \delta Q^A_{,\mu}(x) \tag{2.39b}$$

with

$$\delta Q^A(x) = \varepsilon^a \left[T_a\right]^A_B Q^B(x) \tag{2.40}$$

and analogously for $Q^A_{,\mu}(x)$. Here ε^a; $a = 1,2, \ldots n$, is a set of n <u>constant</u> infinitesimal parameters, and the N x N matrices T_a provide a representation of the Lie

---

x) The system described by $\mathcal{L}^{(0)}(x)$ seems to correspond in the geometric formalism to the specialization that the bundle reduces to a trivial bundle admitting a flat connexion (see chapter III).

xx) Upper and lower indices a,b,c, ... will automatically be summed from 1 to n (n = order of the group G); while upper and lower indices A,B,C, ... will automatically be summed from 1 to N (N = dimension of the representation of $Q^A(x)$).

algebra $\mathfrak{g}$ of the group G with the commutation relations given by

$$[T_a, T_b] = C^c_{ab} T_c \tag{2.41}$$

where the structure constants C^c_{ab} being antisymmetric in a and b and obeying the Jacobi identities, i.e.

$$C^c_{ab} = -C^c_{ba} \tag{2.42}$$

$$C^d_{ab} C^e_{dc} + C^d_{bc} C^e_{da} + C^d_{ca} C^e_{db} = 0. \tag{2.43}$$

Thus we write the Lagrangian density $\mathcal{L}^{(0)}(x)$ as

$$\mathcal{L}^{(0)}(x) = \mathcal{L}^{(0)}(Q^A(x), Q^A_{,\mu}(x)) \tag{2.44}$$

and assume that the equations of motion for the fields $Q^A(x)$ follow from the variational principle (2.5) with $\mathcal{L}^{(0)}(x)$ given by (2.44). The Euler-Lagrange partial differential equations corresponding to the variational principle are (leaving out the arguments x for simplicity)

$$\frac{\partial \mathcal{L}^{(0)}}{\partial Q^A} - \frac{\partial}{\partial x^\mu}\left(\frac{\partial \mathcal{L}^{(0)}}{\partial Q^A_{,\mu}}\right) = 0. \tag{2.45}$$

The invariance of (2.44) under the transformations (2.39) now implies that

$$\delta \mathcal{L}^{(0)} = \frac{\partial \mathcal{L}^{(0)}}{\partial Q^A} \delta Q^A + \frac{\partial \mathcal{L}^{(0)}}{\partial Q^A_{,\mu}} \delta Q^A_{,\mu} = 0 \tag{2.46}$$

where δQ^A is given by eq. (2.40) and similarly for $\delta Q^A_{,\mu}$. Using in eq. (2.46) the equations of motion (2.45) and the fact that the ε^a are independent parameters it is easy to see that these equations correspond to the n conservation laws

$$\partial_\mu j^\mu_a(x) = 0 \quad ; \quad a = 1, 2, \ldots n \tag{2.47}$$

with the n four-vector currents, $j^\mu_a(x)$, associated with the group G being given by

$$j^{\mu}_{a}(x) = \frac{\partial \mathcal{L}^{(0)}(x)}{\partial Q^{A}_{,\mu}(x)} \left[T_a\right]^{A}_{B} Q^{B}(x). \qquad (2.48)$$

For the free U(1) invariant theory eq. (2.48) reduces to the Dirac current with eq. (2.47) expressing charge conservation.

Let us now extend the invariance group G of $\mathcal{L}^{(0)}(x)$ to the G-gauge invariance of a new Lagrangian $\mathcal{L}'(x)$ containing the old fields $Q^A(x)$ and their derivatives as well as a set of gauge potentials, which we call $A^{a}_{\mu}(x)$, and demand invariance of $\mathcal{L}'(x)$ under

$$Q^{A}(x) \longrightarrow Q'^{A}(x) = Q^{A}(x) + \delta' Q^{A}(x) \qquad (2.49a)$$

$$Q^{A}_{,\mu}(x) \longrightarrow Q'^{A}_{,\mu}(x) = Q^{A}_{,\mu}(x) + \delta' Q^{A}_{,\mu}(x) \qquad (2.49b)$$

with the infinitesimal changes induced by the x-dependent transformation of G — which we denote by δ' — being given by

$$\delta' Q^{A}(x) = \varepsilon^{a}(x) \left[T_a\right]^{A}_{B} Q^{B}(x) \qquad (2.50)$$

and

$$\delta' Q^{A}_{,\mu}(x) = \varepsilon^{a}(x) \left[T_a\right]^{A}_{B} Q^{B}_{,\mu}(x) + \partial_\mu \varepsilon^{a}(x) \left[T_a\right]^{A}_{B} Q^{B}(x) \qquad (2.51)$$

where eq. (2.51) follows from eq. (2.50) since $\delta' Q^{A}_{,\mu}(x) = \partial_\mu\left(\delta' Q^{A}(x)\right)$. In addition to eqs. (2.50) and (2.51) one has a transformation formula for the fields $A^{a}_{\mu}(x)$ yet to be determined. To obtain an idea of the form of $\delta' A^{a}_{\mu}(x)$ we first compute the variation $\delta' \mathcal{L}^{(0)}(x)$ induced by the transformations (2.49) and see, using (2.45), that $\delta' \mathcal{L}^{(0)}(x)$ is no longer zero but proportional to $\partial_\mu \varepsilon^{a}(x)$. In fact, one finds

$$\delta' \mathcal{L}^{(0)}(x) = \frac{\partial \mathcal{L}^{(0)}(x)}{\partial Q^{A}_{,\mu}(x)} \left[T_a\right]^{A}_{B} Q^{B}(x)\, \partial_\mu \varepsilon^{a}(x) \qquad (2.52)$$

This suggests that in order to obtain G-gauge invariance of the Lagrangian $\mathcal{L}'(x)$ including the gauge fields these new fields $A^{a}_{\mu}(x)$ should transform inhomogenously under gauge transformations such that the term proportional to $\partial_\mu \varepsilon^{a}(x)$ in eq. (2.52) can be compensated and one, indeed, obtains the invariance

$$\delta' \mathcal{L}'(x) = 0 \tag{2.53}$$

with

$$\mathcal{L}'(x) = \mathcal{L}'(Q^A(x), Q^A_{,\mu}(x), A^a_\mu(x)) \tag{2.54}$$

together with a transformation rule for the fields $A^a_\mu(x)$ given by the ansatz

$$\delta' A^b_\mu(x) = \varepsilon^a(x)[M_a]^b_c A^c_\mu(x) + \partial_\mu \varepsilon^b(x) \tag{2.55}$$

with still unknown coefficients $[M_a]^b_c$. Computing now $\delta'\mathcal{L}'(x)$ and inserting eqs. (2.50), (2.51) and (2.55), and separating, furthermore, the resulting expression into terms proportional to $\varepsilon^a(x)$ and $\partial_\mu \varepsilon^a(x)$, the coefficients of which should be zero according to eq. (2.53) independently from one another x), one deduces from the terms proportional to $\partial_\mu \varepsilon^a(x)$ that the derivative of the $Q^A(x)$ can only appear in $\mathcal{L}'(x)$ in the form

$$D_\mu Q^A(x) = Q^A_{,\mu}(x) - A^a_\mu(x)[T_a]^A_B Q^B(x) \tag{2.56}$$

and one concludes from the terms proprotional to $\varepsilon^a(x)$ using, moreover, eqs. (2.41) and (2.46) that the matrix elements $[M_a]^b_c$ appearing in eq. (2.55) are given by the structure constants of the group G, i.e.

$$[M_a]^b_c = C^b_{ac}. \tag{2.57}$$

This means that the homogenous part of the transformation rule (2.55) characterizes the potentials $A^a_\mu(x)$ as transforming according to the adjoint representation xx) of the group G. One thus obtains finally the following transformation formulae for

---

x) This is so since $\varepsilon^a(x)$ and $\partial_\mu \varepsilon^a(x)$ can be choosen independently from each other for arbitrary a and μ.

xx) In the adjoint representation the commutation relations of the group G, represented in terms of the matrices M_a with matrix elements (2.57), are provided by the Jacobi identities (2.43).

the $A_\mu^a(x)$ under infinitesimal gauge transformations:

$$A_\mu^b(x) \longrightarrow A_\mu'^b(x) = A_\mu^b(x) + \delta' A_\mu^b(x) \tag{2.58}$$

with

$$\delta' A_\mu^b(x) = \varepsilon^a(x)\, C_{ac}^b A_\mu^c(x) + \partial_\mu \varepsilon^b(x) . \tag{2.55'}$$

It is implicit in the presented deduction that the differentiation process D_μ defined by eq. (2.56) defines a G-gauge invariant derivative with the quantities $D_\mu Q^A(x)$ possessing the same gauge transformation character as have the $Q^A(x)$, i.e.

$$\delta' D_\mu Q^A(x) = \varepsilon^a(x) \left[T_a\right]^A_B D_\mu Q^B(x) . \tag{2.59}$$

Using eqs. (2.56), (2.55') and (2.50) together with the commutation rules (2.41) the transformation formula (2.59) can easily be established directly.

From the discussion given so far it is appearent that the Lagrangian $\mathcal{L}'(x)$ is obtained from the Lagrangian $\mathcal{L}^{(0)}(x)$ by the following replacement analogous to the "minimal replacement" in the U(1) gauge theory treated previously:

$$\partial_\mu \longrightarrow D_\mu = \partial_\mu - A_\mu^a(x)\, T_a . \tag{2.60}$$

For later reference we shall write the matrix operator $D = dx^\mu D_\mu$ in the form

$$D = d + i\,\Gamma(x) \tag{2.61}$$

with $d = dx^\mu \partial_\mu$ and the matrix valued one-form $\Gamma(x)$ given by

$$\Gamma(x) = dx^\mu \Gamma_\mu(x) = i\, dx^\mu A_\mu^a(x)\, T_a . \tag{2.62}$$

After these remarks one can now express the Lagrangian (2.54) in the form

$$\mathcal{L}'(Q^A(x), \partial_\mu Q^A(x), A_\mu^a(x)) = \mathcal{L}^{(0)}(Q^A(x), D_\mu Q^A(x)) . \tag{2.63}$$

In order to construct the full Lagrangian $\mathcal{L}(x)$ containing, besides the interaction between the A- and Q-fields described by $\mathcal{L}'(x)$, also a term corresponding to the free gauge field strengths analogous to the $F_{\mu\nu}(x)$ of electromagnetism we first consider the commutator of two gauge invariant derivatives (2.60) of the fields

$Q(x)$ considered as a column matrix

$$Q(x) = \begin{pmatrix} Q^1(x) \\ \vdots \\ Q^N(x) \end{pmatrix} . \tag{2.64}$$

One finds

$$[\mathcal{D}_\mu, \mathcal{D}_\nu]\, Q(x) = i\, \mathcal{R}_{\mu\nu}(x)\, Q(x) \tag{2.65}$$

with

$$\mathcal{R}_{\mu\nu}(x) = \partial_\mu \Gamma_\nu(x) - \partial_\nu \Gamma_\mu(x) + i\,[\Gamma_\mu(x), \Gamma_\nu(x)] . \tag{2.66}$$

Using eq. (2.62) and the commutation relations (2.41) the matrix-valued tensor $\mathcal{R}_{\mu\nu}(x)$ which is antisymmetric in μ and ν can be written as

$$\mathcal{R}_{\mu\nu}(x) = i\, F^a_{\mu\nu}(x)\, T_a \tag{2.67}$$

with

$$F^a_{\mu\nu}(x) = \partial_\mu A^a_\nu - \partial_\nu A^a_\mu - \tfrac{1}{2} C^a_{bc}\left(A^b_\mu A^c_\nu - A^b_\nu A^c_\mu\right) . \tag{2.68}$$

To determine the transformation character of $\mathcal{R}_{\mu\nu}(x)$ under gauge transformations one starts from eq. (2.59) expressing, as mentioned, the fact that x) $\mathcal{D}_\mu Q(x)$ and $Q(x)$ transform in the same way. The same is true for the second gauge invariant derivatives, $\mathcal{D}_\mu \mathcal{D}_\nu Q(x)$, which also transform like Q itself. From this observation and eq. (2.50) one deduces at once the transformation rule

$$\delta' \mathcal{R}_{\mu\nu}(x) = \varepsilon^a(x)\, [T_a, \mathcal{R}_{\mu\nu}(x)] \tag{2.69}$$

which correspond with eqs. (2.67) and (2.41) to the following <u>homogeneous</u> xx) transformation behaviour for the gauge field strengths $F^a_{\mu\nu}(x)$:

---

x) We continue to use the matrix notation (2.64).

xx) In contradistinction to this compare the inhomogeneous transformation rule (2.55') for the $A^a_\mu(x)$ which in matrix form (2.62) reads

$$\delta' \Gamma_\mu(x) = \varepsilon^a(x)\, [T_a, \Gamma_\mu(x)] + i\,(\partial_\mu \varepsilon^a(x))\, T_a .$$

$$\delta' F^{c}_{\mu\nu}(x) = \varepsilon^{a}(x)\, C^{c}_{ab}\, F^{b}_{\mu\nu}(x) \tag{2.70}$$

showing that the fields $F^{a}_{\mu\nu}(x)$ transform as the adjoint representation of the group G.

For the G-gauge invariant Lagrangian, $\mathcal{L}^{(F)}(x)$, of the pure gauge fields we now put in analogy to electromagnetism

$$\mathcal{L}^{(F)}(x) = -\frac{1}{4} F^{a}_{\mu\nu}(x)\, F_{a}^{\mu\nu}(x) \tag{2.71}$$

where $F_{a}^{\mu\nu}(x)$ is defined in terms of the $F^{a}_{\mu\nu}(x)$ of eq. (2.68) by

$$F_{a}^{\mu\nu}(x) = \eta^{\mu\kappa}\, \eta^{\nu\lambda}\, g_{ab}\, F^{b}_{\kappa\lambda}(x) \tag{2.72}$$

with $\eta^{\kappa\lambda}$ = diag (1,-1,-1,-1) being the metric in Minkowski space-time, and with g_{ab} being the Cartan-Killing metric in the group G, i.e.

$$g_{ab} = g_{ba} = C^{c}_{ad}\, C^{d}_{bc}\,. \tag{2.73}$$

The form (2.71) for the Lagrangian of the free gauge fields can be derived rigorously in the framework of the Lagrangian formalism (see ref. 4). The total G-gauge invariant Lagrangian assumes thus finally the form

$$\mathcal{L}(x) = \mathcal{L}'(x) + \mathcal{L}^{(F)}(x) \tag{2.74}$$

with $\mathcal{L}'(x)$ and $\mathcal{L}^{(F)}(x)$ given by eqs. (2.63) and (2.71), respectively. The equations of motion for the G-gauge invariant theory are in analogy to electromagnetism assumed to follow from the variational principle

$$\delta \int_{\Omega} \mathcal{L}(x)\, d^{4}x = 0 \tag{2.75}$$

with $\mathcal{L}(x)$ as given by eq. (2.74). We finally quote the field equations following from eq. (2.75) without presenting its derivation in detail:

$$\frac{\partial \mathcal{L}(x)}{\partial Q^{A}(x)} - \partial_{\mu}\left(\frac{\partial \mathcal{L}(x)}{\partial Q^{A}_{,\mu}(x)}\right) = 0 \tag{2.76}$$

$$\frac{\partial \mathcal{L}(x)}{\partial A_\mu^a(x)} - \partial_\nu \left(\frac{\partial \mathcal{L}(x)}{\partial A_{\mu,\nu}^a(x)} \right) = 0 \qquad (2.77)$$

where $A_{\mu,\nu}^a(x) = \partial_\nu A_\mu^a(x)$. Also the G-gauge invariant theory possesses n conserved four-vector currents (compare eq. (2.14))

$$j_a^\mu(x) = -\frac{\partial \mathcal{L}(x)}{\partial A_\mu^a(x)} \quad , \qquad (2.78)$$

which can also be defined by x) (compare eq. (2.48))

$$j_a^\mu(x) = \frac{\partial \mathcal{L}'(x)}{\partial D_\mu Q^A(x)} [T_a]_B^A Q^B(x) + \frac{\partial \mathcal{L}^{(F)}}{\partial F_{\mu\nu}^b(x)} C_{ac}^b A_\nu^c(x) \quad . \qquad (2.79)$$

The conservation laws for the currents (2.79), i.e. the equations

$$\partial_\mu j_a^\mu(x) = 0 \; ; \quad a = 1, 2, \ldots \; n \, , \qquad (2.80)$$

follow from the gauge invariance of eq. (2.74), expressed by $\delta' \mathcal{L}(x) = 0$, and the equations of motion (2.76) and (2.77). We mention in passing that the currents defined by eq. (2.48) corresponding in the interacting theory to the first term on the right-hand-side of eq. (2.79) would not be conserved quantities if taken alone. There is a gauge potential contribution present in the definition of the currents $j_a^\mu(x)$ for the G-gauge invariant theory which is necessary to restore current conservation according to eqs. (2.80).

A famous example for the formalism described in this section is the Yang-Mills[3] case which is obtained by taking for G the three-parameter isospin group SU(2) with structure constants $C_{ab}^c = C_{abc} = \varepsilon_{abc}$, where ε_{abc} denotes the Levi-Civitta symbol in three dimensions, and by choosing for Q(x) the two component nucleon isospinor $\Psi(x)$.

---

x) To see the equality of the right-hand sides of eqs. (2.78) and (2.79) one writes down $\delta' \mathcal{L}(x) = 0$, uses the field equations (2.76) and (2.77), and separates into contributions proportional to $\varepsilon^a(x)$, $\partial_\mu \varepsilon^a(x)$, and $\partial_\mu \partial_\nu \varepsilon^a(x)$. The contributions proportional to $\varepsilon^a(x)$ yield eqs. (2.80) below, the terms proportional to $\partial_\mu \varepsilon^a(x)$ yield the equality of the definitions (2.78) and (2.79), and the contributions proportional to $\partial_\mu \partial_\nu \varepsilon^a(x)$ just express the fact that the $F_{\mu\nu}^a(x)$ are antisymmetric in μ and ν.

We mention in concluding this section that for comparison with the geometric formulation of gauge theories to be described in the following chapters we did not write explicitly a coupling constant in front of the A_μ^a-terms on the right-hand-side of eqs. (2.56) and (2.60) which would be the analogue of the electric charge appearing in the minimal coupling expression for the U(1) gauge theory discussed before. We consider such a coupling constant characterizing the strength of the interaction between the $A_\mu^a(x)$ and the $Q^A(x)$ fields as absorbed in the fields $A_\mu^a(x)$ possessing the dimension of an inverse length. If one wished one could always replace the $A_\mu^a(x)$ by $g\tilde{A}_\mu^a(x)$ with g being a constant. Then a factor g appears on the right-hand-side of eq. (2.65) as well as in the quadratic terms in the A-fields on the right-hand side of eq. (2.68) with the latter, in the literature on gauge theories, being often referred to as the selfinteraction of the A-fields with a coupling constant g. In general relativity where the Riemann curvature tensor is expressed in a similar manner in terms of the Christoffel symbols as the $F_{\mu\nu}^a(x)$ of eq. (2.68) are here expressed in terms of the gauge potentials $A_\mu^a(x)$ one, however, does not use such a language. We thus follow this latter convention adequate also in a classical gauge theory.

III. ELEMENTS OF DIFFERENTIAL GEOMETRY

III.1 Manifolds, Fiber Bundles

A fundamental assumption implied by the use of differential geometric methods in physics is the hypothesis that the underlying continuum used for the description of physical phenomena is a differentiable manifold. Familiar examples of a definite choice for the geometric stratum on which physics is thought to evolve are the Minkowski space-time of special relativity, the Riemannian space-time V_4 of general relativity, and the U(1) fiber bundle over Minkowski space-time of Section II.1 for the description of electromagnetic effects in the quantum mechanics of electrons. All these generalized spaces are differentiable manifolds. It need not generally be true that all physical phenomena can, indeed, be described in terms of concepts based on the idea of an underlying continuum with manifold structure. However, it is generally assumed that such a hypothesis is reasonable and useful even in the realm of quantum phenomena [x)].

A manifold M_n is, mathematically spoken, a topological space, which is locally similar to a Euclidean space R_n, i.e. can be covered by local coordinate patches which are images of domains in a Euclidean space R_n [xx)]. One calls a local chart, or local system of coordinates, a pair $(\mathcal{U}_i, \phi_i)$ with $\mathcal{U}_i$ being a neighbourhood of M_n and ϕ_i denoting a homeomorphism [xxx)] of $\mathcal{U}_i$ onto an open subset of R_n. A local chart associates with every point $x \in \mathcal{U}_i \subset M_n$ an n-tupel of real numbers $(x^1, x^2 \ldots x^n)$ called the coordinates of the point x in this chart. A collection of charts $\{\mathcal{U}_i, \phi_i\}$, with i in some indexing set J, is called an atlas of M_n. A C^r-manifold can then be characterized by the following properties:

---

x) In a letter of Albert Einstein to Erwin Schrödinger from December 22, 1950 (13) Einstein remarks that he would not consider giving up the idea of an underlying continuum as long as there are no really serious reasons against it.

xx) The four-dimensional hyperbolic Riemannian space of general relativity with signature (+, -, -, -) (see below) is an example of a metric manifold locally homeomorphic to Minkowski space being a pseudo-Euclidean space $R_{1,3}$. Thus after the introduction of a Riemannian metric on a manifold M_n one considers the case of M_n being locally homeomorphic to a Euclidean space R_n (Riemannian manifold) or being locally homeomorphic to a pseudo-Euclidean space $R_{p,q}$, p+q=n, (pseudo-Riemamnian manifold).

xxx) A homeomorphism is a continuous one-to-one mapping between topological spaces.

1) The neighbourhoods $\mathcal{U}_i$ of the atlas $\{\mathcal{U}_i, \phi_i\}$ cover M_n.
2) For two local charts $(\mathcal{U}_i, \phi_i)$ and $(\mathcal{U}_j, \phi_j)$, and $x \in \mathcal{U}_i \cap \mathcal{U}_j$ the homeomorphism $\phi_i \circ \phi_j^{-1}$:
$$\phi_j(\mathcal{U}_i \cap \mathcal{U}_j) \rightarrow \phi_i(\mathcal{U}_i \cap \mathcal{U}_j)$$
defines a C^r map of an open subset of R_n onto an open subset of R_n.

Property 2) states that in the overlap region of two neighbourhoods the coordinates of a point x in one coordinate system are C^r-functions[x)] of the coordinates in the other coordinate system. The Jacobian of the local coordinate transformation is, moreover, assumed to be nonvanishing. Stated more generally we shall assume that the manifold is orientable i.e. that the Jacobian of the local coordinate transformations has a definite sign in all the intersections of the covering $\{\mathcal{U}_i\}$ of an atlas. In the following we shall not explicitely specify the degree of differentiability r and assume that the coordinate functions are as often continuously differentiable as required for the physical arguments to go through. Furthermore, we shall assume that the manifold is a Hansdorff space and is paracompact[xx)]. For the notion of a function defined on a manifold M_n and the concept of a vector or tensor field defined on M_n as well as the definition of the tangent and cotangent spaces at a point $x \in M_n$ we have to refer to the literature quoted.

A mapping $f : M_n \rightarrow M'_{n'}$, between manifolds M_n and $M'_{n'}$, is called "into" (french "dans") or injective at a point $x \in M_n$ if for all local coordinate systems the Jacobian of the transformation $x'^i = f^i(x^1, x^2 \ldots x^n)$; i = 1, ... n', has rank s=n with $n \leq n'$; it is called "onto" (french "sur") or surjective at $x \in M_n$ if the Jacobian has rank s=n' with $n \geq n'$. A diffeomorphism is a one-to-one mapping f from M_n to M'_n for which f and f^{-1} are differentiable. Thus a diffeomorphism is both injective and surjective (s=n=n').

We now turn to a definition of a fiber bundle which is a generalized space possessing manifold structure being locally the topological product of two spaces, the base space and the fiber. A fiber bundle E=E (B,F,π,G,ϕ) over a base manifold B is given by the following collection of objects:

---

x) I.e. r times continuously differentiable functions.

xx) As a short general introduction to differential geometric concepts as used, for example, in general relativity see the first chapter of the book by S.W. Hawking and G.F.R. Ellis(14). As general references on differential geometry and fiber bundles we quote the books by K. Nomizu(15), S. Kobayashi and Nomizu(16), A. Lichnerowicz(17) and Y. Choquet-Bruhat(18).

1) A manifold E called the bundle space or the fiber bundle.
2) A manifold B called the base space.
3) A topological space F, called the fiber [x)].
4) A surjective mapping π of E onto B called the projection. For $x \in B$ one calls $\pi^{-1}(x) = F_x$ the "fiber over x".
5) A covering of B with neighborhoods $\{\mathcal{U}_j\}$ such that $\pi^{-1}(\mathcal{U}_j)$ is homeomorphic to the topological product $\mathcal{U}_j \times F$ under a homeomorphism $\phi_{\mathcal{U}_j}$ such that for $p = (x, \xi_x) \in E$ with $x \in \mathcal{U}_j$ and $\xi_x \in F_x$ one has $\phi_{\mathcal{U}_j}(p) = (x, \bar{\phi}_{\mathcal{U}_j}(\xi_x))$ with $\bar{\phi}_{\mathcal{U}_j}$ denoting a homeomorphism of F_x onto F.
6) A topological group [xx)] G of homeomorphisms of F onto itself, called the structural group of the bundle, such that for $x \in \mathcal{U}_i \cap \mathcal{U}_j$ the homeomorphism $\bar{\phi}_{\mathcal{U}_i} \circ \bar{\phi}^{-1}_{\mathcal{U}_j}$ is an element g of G which depends continuously on x.

The family of mappings $\psi_{ij} = \bar{\phi}_{\mathcal{U}_i} \circ \bar{\phi}^{-1}_{\mathcal{U}_j}$ are called the <u>transition functions</u> of the bundle E corresponding to the covering $\{\mathcal{U}_j\}$ of B. The property 5) states that any bundle is <u>locally trivial</u>. A trivial fiber bundle is a bundle which is globally a direct product of base space and fiber i.e. E = B x F. In this case B can be covered by just one neighbourhood U and the group G reduces to the identity element g=e. A nontrivial fiber bundle is thus a generalization of a direct product space by replacing the topological product by a family of homeomorphisms of $F_x \to F$ such that two homeomorphisms differ by an element of a group G[xxx)]. As will be seen this is exactly the geometric construction which is needed in physics for the formulation of gauge theories.

A family φ_g, $g \in G$, of diffeomorphisms of a differentiable manifold M_n is called a Lie group of differentiable transformations of M_n if G is a Lie group and the mapping of $G \times M_n$ into M_n defined by

$$(g, x) \longrightarrow \varphi_g(x);\ g \in G;\ x \in M_n,\ \varphi_g(x) \in M_n$$

has the following properties:

1) it is differentiable,
2) $\varphi_{g_1 g_2} = \varphi_{g_1} \circ \varphi_{g_2}$
3) φ_e is the identity transformation.

---

x) Also called the "standard fiber" or, in the french literature, the "fibre type".

xx) In the following G will always be a Lie group, i.e. a topological group which is a C^∞- manifold such that $\bar{\phi}_{\mathcal{U}_i} \circ \bar{\phi}^{-1}_{\mathcal{U}_j}$ (see below) is a diffeomorphism depending differentiably on x in $\mathcal{U}_i \cap \mathcal{U}_j$.

xxx) Sloppily one could say that a fiber bundle is a direct product of base space B and fiber F <u>modulo</u> the action of a group G operating on F.

It follows from 2) and 3) that $\varphi_{g^{-1}} = (\varphi_g)^{-1}$. G is said to operate effectively on M_n if φ_g is the identity transformation on M_n only for g=e.

On a Lie group G two Lie groups of diffeomorphisms can be defined in the following way:

$$L_g g_1 = g g_1 \quad , g \in G ; g_1 \in G \tag{3.1a}$$

$$R_g g_1 = g_1 g \quad , g \in G ; g_1 \in G \tag{3.1b}$$

Eq. (3.1a) defines the so-called left translation, and eq. (3.1b) the so-called right translation of an element g_1 in the group G.

If the structural group G of a bundle is a Lie group of diffeomorphisms of F with G being the same manifold as F itself, i.e. if G acts on itself by left (or right) translation in the group, then one calls the bundle $P(B,G,\pi_P,\phi)$, or simply P(B,G), a <u>principal fiber bundle</u> over B.

In the literature a fiber bundle E over B with standard fiber F is frequently referred to as a bundle $E(B,F,\pi_E,G,P,\phi')$ <u>associated</u> with the principal fiber bundle $P(B,G,\pi_P,\phi)$ in the following way x) (16,19). Let F be a manifold on which G acts effectively as a transformation group. If E is identified with the coset space K = PxF/G with P being the mentioned principal fiber bundle over B and π_E being the mapping of K onto B induced by the mapping of π_P of P onto B then one can construct a family of homeomorphisms $\{\phi'_{U_i}\}$ of U_i x F onto $\pi_E^{-1}(U_i)$ xx). In this terminology the tangent bundle over a manifold M_n is the bundle with standard fiber R_n associated with $L(M_n)$, the bundle of linear frames over M_n. These fiber bundles are defined in the following way:

The <u>bundle of linear frames</u> of a manifold M_n of dimension n is the union of the spaces $L_x(M_n)$ of all linear frames λ_x with origin x for all points $x \in M_n$, i.e.

$$L(M_n) = \bigcup_{x \in M_n} L_x(M_n) \tag{3.2}$$

x) Principal fiber bundles play a fundamental role in differential geometry since connexions can only be defined on them (see below). A connexion in an associated bundle E with standard fiber F, i.e. in a bundle associated with a principal bundle P in the way described, is a connexion induced in E through the connexion defined on P.

xx) An associated bundle E is called a vector bundle over B if F provides a representation space for the group G.

$L(M_n)$ is a principal fiber bundle with structural group $Gl(n,R)$.

The union of all pairs (x, ϑ_x) with $x \in M_n$ and $\vartheta_x \in T_x(M_n)$, where $T_x(M_n)$ denotes the tangent space of M_n at x, can be endowed with the structure of a differentiable manifold called the tangent bundle over M_n which can be written as the union of all tangent spaces of the manifold M_n:

$$T(M_n) = \bigcup_{x \in M_n} T_x(M_n) \quad . \tag{3.3}$$

$T(M_n)$ is a fiber bundle over M_n with standard fiber $F = R_n$ and structural group $Gl(n,R)$.

The dual space to the space of tangent vectors $T_x(M_n)$ at $x \in M_n$ is the space of linear differential forms, or one-forms ϑ^*_x, which is denoted by $T^*_x(M_n)$. In analogy to eq. (3.3) the union

$$T^*(M_n) = \bigcup_{x \in M_n} T^*_x(M_n) \tag{3.4}$$

is called the cotangent bundle.

If

$$e_1, e_2, \dots \quad e_n \tag{3.5}$$

is a basis in $T_x(M_n)$ corresponding to a particular element $\lambda_x \in L_x(M_n)$, and if

$$\theta^1, \theta^2, \dots \quad \theta^n \tag{3.6}$$

is a corresponding basis in $T^*_x(M_n)$, an element $\vartheta_x \in T_x(M_n)$ or $\vartheta^*_x \in T^*_x(M_n)$ can be decomposed, respectively, in the following way[x)]

$$\vartheta_x = \vartheta^i e_i , \tag{3.7a}$$

$$\vartheta^*_x = \vartheta_i \theta^i . \tag{3.7b}$$

The so-called natural frame or natural basis in $T_x(M_n)$ is the frame provided by the differentiations along the local coordinate directions i.e.

---

x) We use the summation convention and sum repeated upper and lower indices from 1 to n.

$$e_i = e_i(x) = \frac{\partial}{\partial x^i} \; ; i = 1,2,\ldots n \; , \quad (3.8)$$

correspondingly, the natural basis in the dual space $T_x^*(M_n)$ is given by the coordinate differentials

$$\theta^i = \theta^i(x) = dx^i \; ; i = 1,2,\ldots n \; . \quad (3.9)$$

The transformation rules for the frame vectors $e_i(x)$ and its duals $\theta^i(x)$ in the intersection region of two local charts of an atlas covering M_n is provided by the formulae

$$e'_k = \frac{\partial x^i}{\partial x'^k} e_i \; , \quad (3.10a)$$

$$\theta'^k = \frac{\partial x'^k}{\partial x^i} \theta^i . \quad (3.10b)$$

An arbitrary frame in $T_x(M_n)$, a so-called moving frame, is determined by

$$\tilde{e}_i(x) = [a^{-1}(x)]^j_{\;i} \frac{\partial}{\partial x^j} \quad (3.11)$$

with the matrix $a(x)$ being an element of Gl(n,R). Correspondingly, a moving frame in $T_x^*(M_n)$ is given by

$$\tilde{\theta}^i(x) = [a(x)]^i_{\;j} \, dx^j . \quad (3.12)$$

In the following we shall frequently omit the tilde and denote by $e_i = e_i(x)$ and $\theta^i = \theta^i(x)$ an arbitrary basis in $T_x(M_n)$ and $T_x^*(M_n)$, respectively, which can be the natural frame (3.8) and (3.9) or a moving frame (3.11) and (3.12).

The discussion given so far allows us to consider an arbitrary tensor bundle of type (p,q) over M_n by taking the direct product of tangent and cotangent spaces, $\otimes^p T_x^*(M_n) \otimes^q T_x(M_n)$, at x and considering all pairs $(x, t_p^q(x))$ with $t_p^q(x)$ denoting a tensor at x which is p-fold covariant and q-fold contravariant being given in local coordinates by

$$t^q_p(x) = t^{j_1 j_2 \dots j_q}_{i_1 i_2 \dots i_p}\, \theta^{i_1} \otimes \dots \otimes \theta^{i_p} \otimes e_{j_1} \otimes \dots \otimes e_{j_q} . \tag{3.13}$$

The resulting structure is a fiber bundle $T^q_p(M_n)$ composed of the following set of objects:

Basis: $B = M_n$

Fiber over x: $F_x = \otimes^p T^*_x(M_n) \otimes^q T_x(M_n)$

homeomorphic to the standard fiber $F = \otimes^p R^*_n \otimes^q R_n$

Structural group: $G = \otimes^{p+q}\ Gl(n,R)$ [x)]

Projection: $\pi : (x, t^q_p(x)) \longrightarrow x$.

As a final example of a fiber bundle we consider the bundle of linear differential forms of degree p over M_n which is identical to the bundle of completely antisymmetric tensors of type (p,0). Let us denote by $\wedge T^*_x(M_n)$ the exterior algebra over $T^*_x(M_n)$. Then a p-form on M_n is an assignment of an element of degree p in $\wedge T^*_x(M_n)$ to every $x \in M_n$ in a differentiable manner on M_n. The fiber bundle $D_p(M_n)$ of differential p-forms is a bundle over M_n with standard fiber $F = R^*_n \wedge R^*_n \wedge \dots \wedge R^*_n$ (p times) and structural group $\otimes^p\ Gl(n,R)$ associated with the bundle of linear frames $L(M_n)$. In terms of local coordinates a p-form on M_n is given by (using, moreover, the natural basis (3.9))

$$\begin{aligned} \omega &= \frac{1}{p!}\, \omega_{i_1 i_2 \dots i_p}(x)\, dx^{i_1} \wedge dx^{i_2} \wedge \dots \wedge dx^{i_p} \\ &= \sum_{i_1 < i_2 \dots < i_p} \omega_{i_1 i_2 \dots i_p}(x)\, dx^{i_1} \wedge dx^{i_2} \wedge \dots \wedge dx^{i_p} \end{aligned} \tag{3.14}$$

with completely anisymmetric tensor components, $\omega_{i_1 i_2 \dots i_p}(x)$, denoting the coefficients of the p-form ω in the local coordinate basis. The p-form ω can be regarded as a cross section on $D_p(M_n)$. Generally a <u>cross section</u> on a fiber bundle bundle E over B is defined as a differentiable mapping σ of B into E,

$$\sigma : \quad \begin{aligned} &B \longrightarrow E \\ &\pi \circ \sigma = 1 \end{aligned} \tag{3.15}$$

,

---

x) More exactly, $g \in G$ acts on F as $g = \otimes^p \bar{a} \otimes^q a$ with $a \in Gl(n,R)$ and $\bar{a} = a^{-1}$ corresponding to the convention adopted that the contravariant components of a vector transform with the matrix a and the covariant components transform with a^{-1}.

with the property that the combined mapping of σ and the projection π is the identity mapping.

A system of coordinate frames chosen in a differentiable fashion on M_n can now be defined as a cross section on the frame bundle $L(M_n)$; i.e. a system of moving frames [x)]

$$\lambda_x = (e_1(x), e_2(x), \dots \quad e_n(x)) \tag{3.16}$$

on M_n is given by a cross section on $L(M_n)$. Analogously, a tensor field on M_n is given by a cross section on $T^q_p(M_n)$ and, in particular, a vector field on M_n is a cross section on the tangent bundle $T^1_o(M_n) = T(M_n)$. In the next chapter we shall introduce a generalized spinorial matter wave function capable of describing degress of freedom associated with the strong interactions which can be defined as a cross section of a certain fiber bundle constructed over space-time as base space with the bundle possessing a structural group G to be associated with the dynamics of strong interaction physics.

In closing this section we briefly discuss metric manifolds. A Riemannian manifold, or a Riemannian space, V_n, is a manifold on which a Riemannian metric is defined i.e. a symmetrical covariant tensorfield g of second degree possessing the following properties:

For every $x \in V_n$ g defines a nondegenerate symmetric bilinear form on $T_x^*(V_n) \otimes T_x^*(V_n)$ which can be expressed in local coordinates as

$$ds^2 = g_{ij}(x)\, \theta^i \otimes \theta^j \tag{3.17}$$

where θ^i ; i = 1 ... n, denotes an arbitrary basis in $T_x^*(V_n)$, and the $g_{ij}(x)$ represent a symmetric matrix with nonzero determinant, i.e. $g_{ij}(x) = g_{ji}(x)$; $\det g_{ij} \neq 0$. The structure defined by g is called <u>properly Riemannian</u> if g is positive definite, i.e. $g(\vartheta_x, \vartheta_x) > 0$ for all $\vartheta_x \in T_x(V_n)$. Choosing a local basis this reads

$$g_{ij}\, \vartheta^i \vartheta^j > 0 \tag{3.18}$$

for arbitrary vectors ϑ^i of $T_x(V_n)$.

---

x) We have left out again the tilde written in eq. (3.11).

The definition of a pseudo-Riemannian manifold and the concept of signature will be mentioned below.

A Riemannian metric g allows the definition of a scalar product of two vectors in $T_x(V_n)$

$$\langle u, v\rangle = \langle v, u\rangle = g_{ij}\, u^i v^j \tag{3.19}$$

implying the definition of a positive definite norm $\langle v, v\rangle$ of a vector v according to eq. (3.18). In a certain basis where $u = u^i e_i$ and $v = v^i e_i$ eq. (3.19) implies that

$$\langle e_i, e_j\rangle = g_{ij}. \tag{3.20}$$

The real symmetric matrix $g_{ij}(x)$ can at each point $x \in V_n$ be transformed to diagonal form (principal axes transformation). Writing down the original expression (3.17) in a natural basis one has

$$ds^2 = g_{ij}\, dx^i \otimes dx^j = \sum_{i=1}^{n} (\theta^i)^2 = \theta^i \otimes \theta^j \delta_{ij} \tag{3.21}$$

with $\theta^i(x) = [a(x)]^i_j\, dx^j$ being an x-dependent transformation of Gl(n,R) (compare eq. (3.12)) transforming from the natural (oblique) frame to a particular (namely orthogonal) moving frame denoted by $\theta^i(x)$. The corresponding base vectors of a moving frame in $T_x(V_n)$ are given by

$$e_i(x) = [a^{-1}(x)]^j_i\, \partial_j \quad . \tag{3.11'}$$

Since V_n is endowed with a scalar product one can now, as already mentioned, demand that the frame vectors e_i are orthonormal i.e.

$$\langle e_i, e_j\rangle = \delta_{ij} \quad . \tag{3.22}$$

This together with eqs. (3.11') and (3.20), the latter written down for an oblique natural basis, implies that

$$g_{kl}(x)\, [a^{-1}(x)]^k_i\, [a^{-1}(x)]^l_j = \delta_{ij} \tag{3.23}$$

which just expresses the possibility of a principal axes transformation mentioned above. The question appearing now is whether the frame defined by eqs. (3.11') is, indeed, unique. The answer is negative. A whole family of frames $e'_i(x)$ related by transformations of the group SO(n), i.e. the subgroup of orthogonal n x n matrices of Gl(n,R) with determinant +1 obeying

$$A^k_i(x)\, A^l_j(x)\, \delta_{kl} = \delta_{ij} \tag{3.24}$$

satisfies the above requirements. That is, using the orthogonal frame

$$e'_i(x) = [A^{-1}(x)]^j_i\, e_j(x) \tag{3.25}$$

obtained by transforming with $[a'^{-1}(x)]^k_j = [a^{-1}(x)]^k_i\, [A^{-1}(x)]^i_j$ from the natural basis would clearly also satisfy eq. (3.23) because of (3.24). Given thus the metric on V_n one has still the freedom of choosing from a family of orthonormal moving frames, $e_i(x)$, a particular one at each point x in a smooth manner on V_n. The family of all orthonormal frames with origine x defines a linear space isomorphic to the group SO(n) since any frame can be obtained from a particular one by the action of an element A(x) of SO(n)(compare eq. (3.25)). Considering all orthonormal frames at all the points x of a properly Riemannian manifold V_n leads to the bundle of orthonormal frames $O(V_n)$ over V_n being a principal fiber bundle over V_n with structural group SO(n). A system of orthonormal moving frames on a V_n can thus be given by a cross section on $O(V_n)$.

Before we turn to pseudo-Riemannian manifolds we mention that the scalar product (3.19) allows the definition of a canonical relation between a covariant vector v^i (with respect to a basis e_i in $T_x(V_n)$) and a linear differential form, i.e. an element of $T^*_x(V_n)$ having components (with respect to the dual basis)

$$v_i = g_{ij}\, v^j \quad . \tag{3.26}$$

Usually one identifies these vectors and says that v^i are the contravariant components and v_i the covariant components of the same abstract vector possessing components canonically related to each other according to eq. (3.26). The correspondence (3.26) can be inverted since the metric is nondegenerate, i.e. the g_{ij} possess an inverse. One writes

$$v^i = g^{ij}\, v_j \quad . \tag{3.26'}$$

In the case the fundamental form (3.17) is not positive definite, corresponding to the left-hand-side of eq. (3.18) being larger than zero, zero, or smaller than zero depending on the choice of v^i, the principal axis transformation discussed above yields the form

$$ds^2 = g_{ij}(x)\, dx^i \otimes dx^j = \sum_{i=1}^{s} (\theta^i)^2 - \sum_{i=s+1}^{n} (\theta^i)^2 \tag{3.27}$$

with the systems of signs $(\underbrace{++ \dots +}_{s \text{ times}} \underbrace{-- \dots -}_{(n-s) \text{ times}})$ required to be the same for all $x \in V_n$. In this case one speaks of a pseudo-Riemannian space or pseudo-Riemannian manifold with signature s. A particularly interesting class of pseudo-Riemannian spaces are the hyperbolic spaces with s = 1 (or s = n - 1) i.e. with the system of signs in eq. (3.27) given by (+, -- ... -). Let us illustrate the physically particularly interesting case of a hyperbolic V_4 being the pseudo-Riemannian space-time of general relativity.

We label in this case the natural basis, following common usage, by a greek index running over 0,1,2,3 (global index) and the index of the local frame vectors — now constituting a local Lorentz frame — by a latin index (local Lorentz index). The transformation to a local Lorentz basis than reads in analogy to eq. (3.11')

$$e_i(x) = \lambda^\mu_i(x) \frac{\partial}{\partial x^\mu} \quad ; \; i = 0,1,2,3 \tag{3.28}$$

and, correspondingly, in the dual tangent space at x

$$\theta^i(x) = \lambda^i_\mu(x)\, dx^\mu \quad ; \; i = 0,1,2,3 \; . \tag{3.29}$$

The sixteen fields $\lambda^\mu_i(x)$ are called the vierbein or tetrad fields being matrix elements of a transformation $\lambda(x) \in Gl(4,R)$ from an oblique basis $e_\mu = \partial_\mu$ to an orthonormal basis e_i in $T_x(V_4)$. The inverse transformation, given by $\lambda^i_\mu(x)$, describes the corresponding transformation in $T_x^*(V_4)$ $(\lambda^i_\nu(x)\lambda^\mu_i(x) = \delta^\mu_\nu)$. Because of the orthogonality relations for a time-like basis vector $e_0(x)$ and three space-like basis vectors $e_{1,2,3}(x)$ given by

$$\langle e_i, e_j \rangle = \eta_{ij} \tag{3.30}$$

with η_{ij} = diag(1,-1,-1,-1) the relations (3.23) now read

$$g_{\mu\nu}(x)\,\lambda^{\mu}_{i}(x)\,\lambda^{\nu}_{j}(x) = \eta_{ij} \quad . \tag{3.31}$$

This is unique except for an arbitrary x-dependent Lorentz rotation affecting the latin index corresponding to the transition to a rotated Lorentz frame

$$e'_{i}(x) = \left[\Lambda^{-1}(x)\right]^{j}_{i}\, e_{j}(x) \tag{3.32a}$$

and

$$\theta'^{i}(x) = \left[\Lambda(x)\right]^{i}_{j}\, \theta^{j}(x) \tag{3.32b}$$

in $T_x(V_4)$ and $T^*_x(V_4)$, respectively, with $\Lambda(x) \in O(3,1)^{++}$. Since the space-time manifold V_4 is assumed to be space and time orientable one has to restrict the transformations $\Lambda(x)$ to the subgroup of proper (detΛ = +1) orthochronons ($\Lambda^0_0 > 1$) Lorentz transformations dentoed by $O(3,1)^{++}$ with the first + sign referring to sign $\Lambda^0_0 > 0$ and the second one referring to sign det$\Lambda > 0$. Eqs. (3.31) are invariant under local Lorentz rotations since

$$\Lambda^{k}_{i}(x)\,\Lambda^{l}_{j}(x)\,\eta_{kl} = \eta_{ij} \quad . \tag{3.33}$$

In a V_4 the symmetric metric tensor $g_{\mu\nu}(x)$ possesses ten independent components at each space-time point x. The vierbein fields $\lambda^{\mu}_{i}(x)$, on the other hand, represent sixteen fields. There are thus at each space-time point six additional degrees of freedom above those described by the $g_{\mu\nu}(x)$. These just correspond to the freedom of choosing a particular Lorentz frame from the six-parameter family of all local Lorentz frames at x. Denoting the bundle of Lorentz frames over a hyperbolic V_4 by $L(V_4)$, being a principal fiber bundle over space-time with structural group $O(3,1)^{++}$, one can define what could be called a Weyl-gauge on V_4 [20], namely a smooth system of tetrads on V_4, as a cross section on the Lorentz frame bundle $L(V_4)$. We shall not study the difficult mathematical question under what conditions such a cross section or global system of Lorentz frames exists on space-time. On physical grounds we shall assume that a cross section exists not only locally - which is evident - but also globally. A global cross section on a bundle can quite generally be visualized as a collection of local cross sections given over each local chart of an atlas for the base manifold together with gauge transformations in the intersection regions of the local charts depending differentiably on x.

To discuss spinor fields on a curved space-time manifold it is not sufficient to define the Lorentz frame bundle over space-time and a cross section therein representing a differentiable system of reference frames on V_4. Since the basic two-component spinors transform under the covering group SL(2,C) of the Lorentz group one has to go over to a bundle with structural group SL(2,C). We call a spinor structure or spinor bundle over space-time a fiber bundle $S(V_4, F=C_2, G=SL(2,C), L)$ associated to $L(V_4)$ possessing as fiber a two-dimensional complex space C_2 serving as the representation space for the basic (say undotted) two-component spinors of the group SL(2,C) with SL(2,C) being the structural group of the bundle S. The group homomorphism between SL(2,C) and $O(3,1)^{++}$ (the latter being, as mentioned, the structural group of $L(V_4)$) is expressed by

$$\sigma^k \, [\Lambda^{-1}(x)]^{j}_{\;k} = \tilde{D}^{(\frac{1}{2},0)}(\Lambda(x))\, \sigma^j \,[D^{(\frac{1}{2},0)}(\Lambda(x))]^{-1} . \tag{3.34}$$

Here $\sigma^k = (\sigma^0 = 1, \sigma^s)$, with σ^s ; s = 1,2,3 denoting the Pauli matrices $\left(\begin{pmatrix}0&1\\1&0\end{pmatrix}, \begin{pmatrix}0&-i\\i&0\end{pmatrix}, \begin{pmatrix}1&0\\0&-1\end{pmatrix}\right)$, and $D^{(\frac{1}{2},0)}(\Lambda) \in SL(2,C)$ with $\tilde{D}^{(\frac{1}{2},0)}(\Lambda) = [D^{(\frac{1}{2},0)}(\Lambda^{-1})]^{\dagger} = D^{(0,\frac{1}{2})}(\Lambda)$ where $D^{(\frac{1}{2},0)}(\Lambda)$ and $D^{(0,\frac{1}{2})}(\Lambda)$ are the basic non-equivalent 2x2 representation matrices of SL(2,C) in the standard notation (see, for example Carruthers[21]). It was shown by Geroch[22] that a spinor structure exists on V_4 when a global system of tetrads exists on V_4.

Similarly to what has been said about two-component spinors a Dirac spinor structure of four component type is given by the following spinor bundle associated with the bundle of Lorentz frames $L(V_4)$ over space-time

$$S(V_4, F=C_4, G=SL(2,C)\oplus SL(2,C)^{*}, L) \tag{3.35}$$

with C_4 being a complex linear space on which the group $SL(2,C)\oplus SL(2,C)^{*}$ acts as a transformation group. The connections between the four dimensional spinor representation of the orthochronous Lorentz group and the group $O(3,1)^{++}$ is provided by the well-known formula analogous to eq. (3.34),

$$\gamma^k \, [\Lambda^{-1}(x)]^{j}_{\;k} = S(x)\, \gamma^j \, S^{-1}(x) \tag{3.36}$$

with γ^k; k=0,1,2,3 being the four Dirac matrices obeying the relations (2.7), however, now with a local latin index, and S(x) being given in the so-called γ^5-diagonal representation by

$$S(x) = \begin{pmatrix} D^{(\frac{1}{2},0)}(\Lambda(x)) & 0 \\ 0 & D^{(0,\frac{1}{2})}(\Lambda(x)) \end{pmatrix} \tag{3.37}$$

with the basic nonequivalent two-dimensional representation matrices of SL(2,C) as introduced before.

A four component Dirac spinor field $\Psi(x)$ on a curved space-time manifold V_4 can now be defined as a cross section of the spinor bundle (3.35) which we write as

$$\Psi(x) = \psi(x, e_i(x)) \quad . \tag{3.38}$$

We denote by $\Psi(x)$ the spinor field <u>in abstracto</u> and by $\psi(x, e_i(x))$ its <u>representative in a certain Weyl gauge on V_4</u>, i.e. as given with respect to a definite system of moving orthonormal Lorentz frames, $e_i(x)$, on V_4 determined, as mentioned, by a cross section of the bundle $L(V_4)$. This is completely analogous to the definition of, for example, a vector field $v(x)$ on space-time defined as a cross section on $T(V_4)$ with a local coordinate representation given by $v(x) = v^i(x) e_i(x)$ where $v^i(x)$ is the representative of $v(x)$ with respect to a system of axes chosen at each point. Changing the cross section on $L(V_4)$ relative to which ψ is measured, i.e. performing a Lorentz gauge transformation which is an x-dependent Lorentz transformation of the local frame in each tangent space for all x on V_4, corresponds to the following gauge transformation of the representative $\psi(x, e_i(x))$ of the spinor field [x)]

$$\psi'(x, e'_i(x)) = S(x)\,\psi(x, e_i(x)) \tag{3.39}$$

where $e'_i(x)$ and $e_i(x)$ are related according to eq. (3.32a), and $\Lambda(x)$ and $S(x)$ are connected by eq. (3.36) defining the homomorphism $SL(2,C) \oplus SL(2,C)^* \longrightarrow O(3,1)^{++}$.

In order to define a Lorentz gauge invariant - or $S(x)$ invariant - differentiation process for a four-component spinor quantity defined on V_4, which is a differentiation process independent on the particular choice of moving Lorentz frames on V_4, one has to introduce a connexion in V_4 or, more exactly, a spinor connexion. On a Riemannian manifold the so-called linear or affine connexion is given in terms of the metric $g_{\mu\nu}(x)$. A spinor connexion requires the knowledge of the Vierbein fields $X^\mu_i(x)$. In the next section we shall review the general theory of connexions on a principal fiber bundle over an arbitrary manifold M_n and study in more detail as a particularly interesting example the linear connexion which is the connexion in the bundle of linear frames $L(M_n)$.

---

x) Later we shall simply call $\psi(x, e_i(x))$ the spinor field on V_4 and, moreover, leave out the frame $e_i(x)$ in the argument.

III.2 Connexions in a Principal Fiber Bundle

We first turn to the discussion of a linear connexion which is, as mentioned at the end of the last section, the connexion in the bundle of linear frames $L(M_n)$ over the manifold M_n. M_n denotes here at first an arbitrary n-dimensional manifold. Riemannian manifolds will be considered below as a special case.

There are two definitions of a linear connexion. We first treat the historically earlier definition making no reference to fiber bundles and then go on to present the more general modern definition of a so-called infinitesimal connexion (or simply a connexion) in a principal fiber bundle $P(M_n,G)$ over a manifold M_n. For $G=Gl(n,R)$ we then recover from the second definition again that of the linear connexion in the bundle of linear frames.

A. First definition of a linear connexion

A linear connexion in a differentiable manifold M_n is a mapping $t_p^q \rightarrow Dt_p^q$ of the tensor fields of type (p,q) into the tensor fields of type (p+1,q) with the following properties:

1)
$$D\,(t_p^q + s_p^q) = Dt_p^q + Ds_p^q \tag{3.40}$$

2) If f is a differentiable function on M_n and df its differential (which is a covariant vector) then

$$D\,f\,t_p^q = df \otimes t_p^q + f\,Dt_p^q\,. \tag{3.41}$$

Point 2) implies that for functions $Df = df$, i.e. the differentiation defined by D is identical to the ordinary differentiation when applied to a function. Dt_p^q is called the absolute covariant derivative of the tensor t_p^q.

Let us apply the operation D to a vector field $v(x)$ given in a local chart with base vectors $e_i = e_i(x)$ by $v(x) = v^i(x)\,e_i(x)$, i.e. using 2):

$$Dv = D(v^i e_i) = dv^i \otimes e_i + v^i\,De_i\,. \tag{3.42}$$

From this formula it is apparent that it is sufficient to know the absolute derivative of the base vectors e_i in order to compute Dv. Since D raises the degree of covariance by one unit one can expand De_i according to

$$De_i = \Gamma_{ki}^{\;j}\,\theta^k \otimes e_j \tag{3.43}$$

with θ^k being an arbitrary basis in $T_x^*(M_n)$ which is dual to e_i. Combining this with eq. (3.42) one has

$$\mathcal{D}v = (dv^i + \omega^i_j v^j) \otimes e_i \tag{3.44}$$

with ω - called the matrix one-form of the connexion - having matrix elements given by

$$\omega^i_j = \theta^k \Gamma^i_{kj} \, . \tag{3.45}$$

With eq. (3.45) one can give eq. (3.43) the easily memorizable form frequently used by E. Cartan[(23)]

$$\mathcal{D}e_i = \omega^j_i e_j \, . \tag{3.46}$$

With

$$d = \theta^k \partial_k \tag{3.47}$$

where the ∂_k are called the Pfaffian derivatives in case an arbitrary moving frame (3.11) is used as basis [x)], the operation D can finally be written as

$$\mathcal{D} = \theta^k \mathcal{D}_k \tag{3.48}$$

with

$$\mathcal{D}_k v^i = \partial_k v^i + \Gamma^i_{kj} v^j \, . \tag{3.49}$$

The mixed tensor $\mathcal{D}_k v^i$ possessing the components shown on the right-hand side of eq. (3.49) is called the covariant derivative of the contravariant vector field $v(x)$ given in the local chart by its components v^i. It is easy to show from the invariant property of the contracted quantity $u_i v^i$ and the rule Df = df that for a covariant vector field $v^*(x) = v_i(x)\, \theta^i(x)$ one obtains the formula

---

[x)] In the natural basis one has, of course, $d = dx^i \frac{\partial}{\partial x^i}$.

$$D_k v_i = \partial_k v_i - \Gamma_{ki}^{\,j} v_j \quad . \tag{3.50}$$

Quite generally, one can easily determine with the help of the equations

$$D_k e_i = \Gamma_{ki}^{\,j}\, e_j \tag{3.51a}$$

and

$$D_k \theta^i = -\Gamma_{kj}^{\,i}\, \theta^j \tag{3.51b}$$

and the form (3.13) of an arbitrary tensor field in a local basis the general formula for the covariant derivative, $D_k\, t^{j_1 \dots j_q}_{i_1 \dots i_p}$, of a tensor field of type (p,q). We only quote as an example the formula for the covariant derivative of a second order mixed tensor field with local components t^i_j :

$$D_k t^i_j = \partial_k t^i_j - \Gamma_{kj}^{\,\ell}\, t^i_\ell + \Gamma_{k\ell}^{\,i}\, t^\ell_j \quad . \tag{3.52}$$

Furthermore, due to the fact that Dv defined in eq. (3.44) is an invariant under transformations of the local system of axes

$$e'_i = [a^{-1}(x)]^j_i\; e_j \tag{3.53a}$$

$$\theta'^i = [a(x)]^i_j\; \theta^j \tag{3.53b}$$

corresponding to a local change of gauge in a certain neighbourhood $\mathcal{U}_j$ of the covering $\{\mathcal{U}_i\}$ of the manifold M_n with $a(x) \in Gl(n,R)$; or, corresponding to the transformations (3.10) in the intersection of two local charts on M_n. Together with the vector character of $dv^i + \omega^i_j v^j$ one thus at once deduces the following transformation rule for the ω^i_j from eq. (3.44):

$$\omega'^h_k = [a]^h_i\; \omega^i_j\; [a^{-1}]^j_k + [a]^h_i\, d[a^{-1}]^i_k \tag{3.54}$$

In matrix form, writing also the x dependence explicitely, eq. (3.54) can be written compactly as

$$\omega'(x) = a(x)\,\omega(x)\,a^{-1}(x) + a(x)\,d a^{-1}(x) \,. \qquad (3.54')$$

Eqs. (3.54) and (3.54') represent the typical inhomogenous transformation formulae for the connexion form $\omega(x)$ under the gauge transformations (3.53) corresponding, as mentioned, to a transition to another local cross section on the linear frame bundle $L(M_n)$, or corresponding to the x-dependent differentiable transformation of frames induced by the relation of two local systems of coordinates in the intersection region of two local charts on M_n.

In order to be able to compare eq. (3.54') more easily with the formulae presented in Sect. II.2 we consider an infinitesimal gauge transformation (3.53) given by

$$a(x) = 1 + \varepsilon^a(x)\,\tilde{T}_a \qquad (3.55a)$$

and

$$a^{-1}(x) = 1 - \varepsilon^a(x)\,\tilde{T}_a \qquad (3.55b)$$

where we have denoted the generators of Gl(n,R) by $\tilde{T}_a$. Then eq. (3.54') reduces to

$$\omega'(x) = \omega(x) + \delta'\omega(x) \qquad (3.56)$$

with

$$\delta'\omega(x) = \varepsilon^a(x)\,[\tilde{T}_a, \omega(x)] + d\varepsilon^a(x)\,\tilde{T}_a \qquad (3.57)$$

This last equation corresponds to eq. (2.55') (compare also eq. (2.62) and the equation given in the footnote quoted after eq. (2.69)).

Let us for completeness also write down eq. (3.54) for the connexion coefficients $\Gamma^{\,j}_{k\,i}$ (leaving out again the argument x)

$$\Gamma'^{\,k'}_{i'\,j'} = [a]^{k'}_{k}\,[a^{-1}]^{i}_{i'}\,[a^{-1}]^{j}_{j'}\,\Gamma^{\,k}_{i\,j} + [a]^{k'}_{s}\,\partial_{i'}[a^{-1}]^{s}_{j'} \,. \qquad (3.58)$$

In order to characterize the connexion coefficients $\Gamma^{\,j}_{k\,i}$ still further we define the covariant derivative of a vector field $v(x)$ in the direction of a given vector u by

$$D_u v = u^j\,(D_j v^i)\,e_i \qquad (3.59)$$

with $D_j v^i$ as given by eq. (3.49). This shows that the covariant derivative of a vector field $v(x)$ in the direction of the basis vector e_k is given by

$$D_{e_k} v = (D_k v^i)\, e_i \qquad (3.60)$$

or, in components,

$$(D_{e_k} v)^i = \partial_k v^i + \Gamma^i_{kj}\, v^j \quad . \qquad (3.61)$$

From eq. (3.61) finally follows, for $v(x)$ being replaced by e_j :

$$(D_{e_k} e_j)^i = \Gamma^i_{kj} \quad . \qquad (3.62)$$

We now turn to the derivation of Cartan's structural equations characterizing the manifold M_n from the point of view of geometry by a torsion and a curvature two-form derived from the connexion form $\omega(x)$. However, before we do this we have to say a few words about exterior derivatives of forms [x)].

The exterior derivative of forms (denoted by the symbol d) is a mapping from the space of linear differential p-forms over M_n into the space of linear differential (p+1) forms over M_n possessing the following properties:

1) If f is a function (a zero-form) then df is a one-form defined by the differential of f.
2) If $\Omega = \omega + \theta$ then $d\Omega = d\omega + d\theta$.
3) $dd\omega = 0$.
4) If ω is a p-form and θ a q-form then

$$d(\omega \wedge \theta) = d\omega \wedge \theta + (-1)^p\, \omega \wedge d\theta. \qquad (3.63)$$

4') If f is a function and θ a q-form then

$$d(f\theta) = df \wedge \theta + f\, d\theta. \qquad (3.64)$$

The comparison of eq. (3.63) with eq. (3.64) leads to the rule that one only writes the symbol $\wedge$ of exterior multiplication if both factors involved are indeed forms and leaves it out if functions are involved.

---

x) For a transparent account of differential forms see refs. 17 and 18 and also the book by H. Flanders[(24)].

We now turn to the problem of deriving Cartan's structural equations. We first observe that whereas the second ordinary derivatives $\partial^2 f/\partial x^i \partial x^j$ do not depend on the order of differentiation this is not the case for covariant differentiations. Let us compute $D_i(\partial_j f)$, with f being a function on M_n, and consider the following commutator using, moreover, a natural basis since only in this case one has $[\partial_i, \partial_j] = 0$:

$$D_i(\partial_j f) - D_j(\partial_i f) = -\left[\Gamma^k_{ij} - \Gamma^k_{ji}\right] \partial_k f \quad . \tag{3.65}$$

The in i and j antisymmetric part of the connexion coefficients Γ^k_{ij} referring to a natural basis is a mixed tensor of order three with components in a local chart given by

$$S^k_{ij} = \Gamma^k_{ij} - \Gamma^k_{ji} \quad . \tag{3.66}$$

That (3.66) is, indeed, a tensor, i.e. transforms homogeneously under coordinate changes in the intersection regions of local charts (compare eqs. (3.10)) can at once be seen from eq. (3.58) by replacing there

$$[a^{-1}(x)]^k_j \quad \text{by} \quad \frac{\partial x^k}{\partial x'^j}$$

and

$$[a(x)]^k_j \quad \text{by} \quad \frac{\partial x'^k}{\partial x^j} \quad .$$

We now transform eq. (3.66) from the natural basis to a moving frame basis given by eqs. (3.11) and (3.12). Denoting temporarily again the quantities referred to a moving frame by a tilde (as we did in eqs. (3.11) and (3.12)) one obtains from eqs. (3.58) and (3.66)

$$\begin{aligned}\tilde{\Gamma}^l_{mn} - \tilde{\Gamma}^l_{nm} &= [a]^l_k [a^{-1}]^i_m [a^{-1}]^j_n S^k_{ij} + \\ &+ [a]^l_s \left\{ \partial_m [a^{-1}]^s_n - \partial_n [a^{-1}]^s_m \right\}\end{aligned} \tag{3.67}$$

where the first term on the right-hand side is equal to $\tilde{S}^k_{ij}$ because of the tensor character of these quantities. The last two terms involving the Pfaffian derivatives ∂_m and ∂_n in an antisymmetric fashion can be abbreviated by $-\varrho^l_{mn}$ so that eq. (3.67), putting, moreover, ϱ^l_{mn} on the left-hand side, reads

$$\varrho^l_{mn} + \tilde{\Gamma}^l_{mn} - \tilde{\Gamma}^l_{nm} = \tilde{S}^l_{mn} \quad . \tag{3.68}$$

Multiplying both sides of this equation by $\frac{1}{2}\tilde{\theta}^m \wedge \tilde{\theta}^n$, with $\tilde{\theta}^m \wedge \tilde{\theta}^n$ providing a moving frame basis of two-forms, the first term on the left-hand side can easily be shown to be the exterior derivative of $\tilde{\theta}^l$, i.e.

$$d\tilde{\theta}^l = \frac{1}{2} \varrho^l_{mn} \tilde{\theta}^m \wedge \tilde{\theta}^n \tag{3.69}$$

such that, finally, eq. (3.68) with the help of eq. (3.45) takes the form

$$d\tilde{\theta}^l + \tilde{\omega}^l_n \wedge \tilde{\theta}^n = \tilde{\tau}^l \quad . \tag{3.70}$$

Here we have put

$$\tilde{\tau}^l = \frac{1}{2} \tilde{S}^l_{mn} \tilde{\theta}^m \wedge \tilde{\theta}^n \tag{3.71}$$

denoting the torsion two-form of the connexion. Since eq. (3.70) is valid for any basis [x)] we drop the tilde again and write the torsion equations or first group of structural equations of E. Cartan as [xx)]

$$d\theta^i + \omega^i_k \wedge \theta^k = \tau^i \quad . \tag{3.70'}$$

Cartan's second group of structural equations is given by

$$d\omega^i_j + \omega^i_k \wedge \omega^k_j = \Omega^i_j \tag{3.72}$$

where Ω^i_j is the curvature two-form. Also these equations are frame independent, i.e. have the same form in a natural basis and in a moving frame. In a natural basis the two-form Ω^i_j can be expanded as

---

x) Specializing eq. (3.70) to a natural basis $\tilde{\theta}^l = dx^l$ one immediately recovers the equation (3.66) from which we started due to $ddx^l = 0$.

xx) The left-hand-side of eq. (3.70') can be called the <u>exterior covariant derivative</u> of the one-form θ^i, i.e.

$$D\theta^i = d\theta^i + \omega^i_k \wedge \theta^k \quad .$$

$$\Omega^i_j = \tfrac{1}{2} R_{k\ell j}{}^i \, dx^k \wedge dx^\ell \tag{3.73}$$

with $R_{k\ell j}{}^i$ being the curvature tensor. Eq. (3.72) states, using $\omega^i_j = dx^k \Gamma^i_{kj}$, that the $R_{k\ell j}{}^i$ are given in a natural basis by

$$R_{k\ell j}{}^i = \partial_k \Gamma^i_{\ell j} - \partial_\ell \Gamma^i_{kj} + \Gamma^i_{ks}\Gamma^s_{\ell j} - \Gamma^i_{\ell s}\Gamma^s_{kj} \quad . \tag{3.74}$$

One derives eq. (3.72) by considering the commutator of two covariant derivatives applied to, say, a vector field with local components v^i. One finds

$$[D_k, D_\ell]\, v^i = R_{k\ell j}{}^i v^j - S^j_{k\ell} D_j v^i \tag{3.75}$$

where the curvature term on the right-hand side appears here in terms of the connexion coefficients as given by the right-hand side of eq. (3.74) and the torsion term appears as expressed by eq. (3.66). Thus eq. (3.74) is established which in turn implies (3.72).

Let us now specialize the developped formalism to the case of a Riemannian manifold V_n. In this case we have the following restrictions:

1) The torsion vanishes, i.e. $\tau^k = 0 \quad (S^k_{ij} \equiv 0)$.
2) The metric is covariant constant, i.e. $D_k g_{ij} = 0$.

Property 1) implies that the connexion coefficients Γ^k_{ij} in the natural basis are <u>symmetric</u> in the lower indices. Property 2) implies that the scalar product (3.19) is invariant under parallel transfer. Property 2) can be rewritten as

$$\partial_k g_{ij} = \Gamma^\ell_{ki} g_{\ell j} + \Gamma^\ell_{kj} g_{i\ell} \quad , \tag{3.76}$$

or as

$$d g_{ij} = \omega^\ell_i g_{\ell j} + \omega^\ell_j g_{i\ell} \tag{3.76'}$$

valid in any basis (natural or moving frame). Assuming the Γ^ℓ_{ik} in eq. (3.76) to refer to a <u>natural basis</u> one derives by cyclic permutation of the indices the Christoffel form for the connexion coefficients expressed in terms of the metric tensor $g_{ij}(x)$ by

$$\Gamma_{ki}^{\ \ l} = \Gamma_{ik}^{\ \ l} = g^{lj}\,\tfrac{1}{2}\left(\partial_k g_{ij} + \partial_i g_{kj} - \partial_j g_{ki}\right) . \tag{3.77}$$

Assuming, on the other hand, the g_{ij} in eq. (3.76') to refer to an <u>orthonormal moving frame</u> corresponds to

$$g_{ij} = \begin{cases} \delta_{ij} & \text{properly Riemannian } V_n \\ \tilde{\eta}_{ij} & \text{hyperbolic}^{x)}\ V_n \\ \pm\delta_{ij} & \text{pseudo-Riemannian } V_n \end{cases} .$$

Eq. (3.76') implies then that

$$\omega_{ij} = -\omega_{ji} \tag{3.78}$$

with

$$\omega_{ij} = \omega_i^{\ l} g_{lj} . \tag{3.79}$$

For a Riemannian manifold V_n Cartan's first structural equation is simply $D\theta^i = 0$ which, in a natural basis, is trivially satisfied. The curvature equation takes the form

$$d\omega_{ij} + \omega_{is} \wedge \omega_j^{\ s} = \Omega_{ij} = -\Omega_{ji} \tag{3.80}$$

with

$$\Omega_{ij} = \Omega_i^{\ k} g_{kj} . \tag{3.81}$$

Analogously one has

$$R_{klij} = R_{kli}^{\ \ \ \ m} g_{mj} \tag{3.82}$$

and

$$\Gamma_{kij} = \Gamma_{ki}^{\ \ l} g_{lj} . \tag{3.83}$$

Going back to the general, not necessarily Riemannian, case the structural equations (3.70') and (3.72) lead to integrability conditions for the curvature and torsion tensor fields on a manifold M_n which follow easily from the exterior differentiation of these equations. One obtains:

---

x) $\tilde{\eta}_{ij}$ = diag (1,-1,-1, ... -1) .

$$d\tau^j = \Omega^j_{\ k} \wedge \theta^k - \omega^j_{\ k} \wedge \tau^k \qquad (3.84)$$

$$d\Omega^j_{\ i} = \Omega^j_{\ k} \wedge \omega^k_{\ i} - \omega^j_{\ k} \wedge \Omega^k_{\ i} \quad . \qquad (3.85)$$

The second equations are the so-called Bianchi identities for the curvature two-form which can be transparently expressed as the vanishing of the exterior covariant derivative of $\Omega^j_{\ i}$, i.e.

$$D\Omega^j_{\ i} = d\Omega^j_{\ i} + \omega^j_{\ k} \wedge \Omega^k_{\ i} - \omega^k_{\ i} \wedge \Omega^j_{\ k} = 0 \quad . \qquad (3.86)$$

In terms of the curvature tensor (3.74) and the torsion tensor (3.66) this equation reads [x)]

$$D_{\{s} R_{kl\}i}{}^{j} = - S^{\ m}_{\{sk} R_{l\}mi}{}^{j} \qquad (3.87)$$

where the indices in the curly brackes have to be summed over a cyclic permutation, and with $D_s R_{kli}{}^j$ as given by

$$D_s R_{kli}{}^{j} = \partial_s R_{kli}{}^{j} - \Gamma^{k'}_{sk} R_{k'li}{}^{j} - \Gamma^{l'}_{sl} R_{kl'i}{}^{j} - \Gamma^{i'}_{si} R_{kli'}{}^{j} + \Gamma^{j}_{sj'} R_{kli}{}^{j'} \quad . \qquad (3.88)$$

For a Riemannian manifold eq. (3.87) reduces to

$$D_s R_{klij} + D_k R_{lsij} + D_l R_{skij} = 0 \quad . \qquad (3.89)$$

From eq. (3.84) follows for a Riemannian manifold at once the conditions

$$R_{skl}{}^{j} + R_{kls}{}^{j} + R_{lsk}{}^{j} = 0 \quad . \qquad (3.90)$$

---

x) To obtain eq. (3.87) we used the expression for $R_{kli}{}^j$ and $S^{\ j}_{kl}$ as given in terms of the $\Gamma^{\ j}_{kl}$ referring to a natural basis. However, eq. (3.87) involves only tensor quantities and is thus true in any frame.

These relations together with

$$R_{klij} = - R_{lkij} \tag{3.91}$$

implied by eq. (3.74) and

$$R_{klij} = - R_{klji} \tag{3.92}$$

implied by eq. (3.80) leads, with eq. (3.83) to the symmetry

$$R_{klij} = R_{ijkl} \quad . \tag{3.93}$$

Before we turn to the second definition of a connexion we like to specialize the developped formalism to the hyperbolic V_4 of general relativity in order to collect various formulae to be used later. Corresponding to the transformation from a natural frame to a moving Lorentz frame (compare the previous section) one has the relations

$$\partial_i = \lambda^{\mu}_{i}(x) \frac{\partial}{\partial x^{\mu}} \quad ; \; i=0,1,2,3 \quad , \tag{3.28'}$$

and

$$\theta^{i} = \lambda^{i}_{\mu}(x) \, dx^{\mu} \quad ; \; i=0,1,2,3 \, . \tag{3.29'}$$

The $\theta^{i}(x)$ represent a set of four Pfaffian forms and the ∂_i (with local latin index i) are the corresponding Pfaffian derivatives obeying

$$[\partial_i , \partial_j] = - \varrho^{k}_{ij} \, \partial_k \tag{3.94}$$

with

$$\varrho^{k}_{ij}(x) = \lambda^{\mu}_{i}(x) \lambda^{\nu}_{j}(x) \left[\partial_{\mu} \lambda^{k}_{\nu}(x) - \partial_{\nu} \lambda^{k}_{\mu}(x) \right] \quad . \tag{3.95}$$

The connexion form, when referred to a local Lorentz basis, is given by

$$\omega_{ij} = \omega^{k}_{i} \, \eta_{kj} = - \omega_{ji} \tag{3.96}$$

leading, together with eq. (3.45), to the relation

$$\Gamma_{ki}^{\ \ l}\eta_{lj} = \Gamma_{kij} = -\Gamma_{kji} \tag{3.97}$$

for the local (i.e. latin indexed) form of the connexion coefficients. The coefficients Γ_{kij} are called the Ricci rotation coefficients(25). Eq. (3.46) can now be written as

$$D\lambda_i^{\mu}(x) = dx^{\nu}\left[\partial_\nu \lambda_i^{\mu}(x) + \Gamma_{\nu\kappa}^{\mu}(x)\lambda_i^{\kappa}(x)\right] = \omega_i^{j}(x)\lambda_j^{\mu}(x) \,. \tag{3.98}$$

With eqs. (3.45) and (3.29') this is seen to be identical to the transformation formula (3.58) for the connexions coefficients from the greek indexed natural basis to the latin indexed local Lorentz basis.

The curvature tensor of the connexion is given by eqs. (3.74) and (3.83) written with greek indices according to the adopted convention for space-time.

We can now finally characterize the connexion matrix $\omega(x)$ as a one-form with values in the Lie algebra of the Lorentz group being the structural group of the Lorentz frame bundle $L(V_4)$. Hence $\omega(x)$ can be expanded in the following way

$$\omega(x) = \frac{i}{2}\,\omega_{kl}(x)\,R^{kl} \,. \tag{3.99}$$

Here $R^{ik} = -R^{ki}$; i,k = 0,1,2,3, are the six generators of the Lorentz group in their basic 4x4 representation possessing matrix elements given by

$$\left[R^{kl}\right]^i_{\ j} = -i\left[\delta_j^k\delta_m^l - \delta_j^l\delta_m^k\right]\eta^{mi} \,. \tag{3.100}$$

The connexion form $\omega(x)$ - assumed to be given in a certain gauge - corresponds thus, according to eq. (3.99), at each $x \in V_4$ to an infinitesimal Lorentz transformation on the Lorentz frame bundle $L(V_4)$ with x-dependent "parameters" $\omega_{kl}(x) = dx^{\mu}\,\Gamma_{\mu kl}(x) = \theta^i(x)\,\Gamma_{ikl}(x)$ possessing one-form character which implies that they are proportional to the "infinitesimal step" taken on the base space V_4 of the bundle in going from the point $x \in V_4$ to an infinitely nearby point. Hence the name rotation coefficients for the $\Gamma_{ikl}(x)$.

It is now an easy matter to define an invariant differentiation process for a Dirac spinor field $\Psi(x)$ on V_4 represented by $\psi(x, e_i(x))$ (compare eq. (3.38)) which is analogous to the formulae (3.44), (3.49) and (3.50), i.e.

$$D\,\psi(x, e_i(x)) = \left[d + i\,\Gamma(x)\right]\psi(x, e_i(x)) \,. \tag{3.101}$$

Here $\Gamma(x)$ denotes the spinor connexion form being a one-form with values in the Lie algebra of the structural group of the spinor bundle (3.35) associated to the bundle of Lorentz frames over V_4. Using the 4x4 spinor representation of the Lorentz group given by

$$M^{ij} = \frac{i}{4}[\gamma^i, \gamma^j] \tag{3.102}$$

the spinor connexion on a space-time manifold V_4 can be written as

$$\Gamma(x) = \frac{1}{2}\omega_{ij}(x) M^{ij} = \frac{1}{2}\theta^k(x)\Gamma_{kij}(x) M^{ij} \quad . \tag{3.103}$$

Again the matrix one-form $i\Gamma(x)$ - for a definite gauge chosen - can be interpreted as an infinitesimal Lorentz transformation in the local spinor space C_4 of the bundle S for each $x \in V_4$ which is associated with the transition from the space-time point x to a neighbouring space-time point.

For the adjoint Dirac spinor field

$$\bar{\psi}(x, e_i(x)) = \psi^\dagger(x, e_i(x))\gamma^0 \tag{3.104}$$

transforming inversely to $\psi(x, e_i(x))$ under Lorentz gauge transformation, i.e.

$$\bar{\psi}'(x, e_i'(x)) = \bar{\psi}(x, e_i(x))\, S^{-1}(x) \tag{3.105}$$

with

$$S^{-1}(x) = \gamma^0 S^\dagger(x)\gamma^0 \tag{3.106}$$

(compare eq. (3.39)), one obtains in analogy to eq. (3.101) the following rule for covariant differentiation with $\overleftarrow{D}$ operating to the left

$$\bar{\psi}(x, e_i(x))\overleftarrow{D} = d\bar{\psi}(x, e_i(x)) - i\,\bar{\psi}(x, e_i(x))\Gamma(x) . \tag{3.107}$$

The Ricci rotation coefficients are real quantities. Due to the relation

$$M^{ij} = \gamma^0 M^{ij\dagger}\gamma^0 \tag{3.108}$$

for the generators of the Lorentz group in the 4x4 Dirac representation one has an analogous reality relation for the spinor connexion $\Gamma(x)$, i.e.

$$\Gamma(x) = \gamma^0 \Gamma^{\dagger}(x) \gamma^0 \; . \qquad (3.109)$$

We close this section by quoting the gauge transformation property for the spinor connexion $\Gamma(x)$ which is analogous to eq. (3.54'), i.e.

$$\Gamma'(x) = S(x)\,\Gamma(x)\,S^{-1}(x) - i\,S(x)\,d S^{-1}(x) \; . \qquad (3.110)$$

As mentioned above this equation not only describes a change of gauge for the spinor connexion $\Gamma(x)$ in a certain domain $\mathcal{U}_j$ of space-time, but it governes also the transformation of the spinor connexion [being the connexion on the spinor bundle $S(V_4, C_4, SL(2,C) \oplus SL(2,C)^*, L)$ associated to the Lorentz frame bundle $L(V_4)$] in the intersection regions of the local charts of an atlas for V_4. As a consequence of eq. (3.110) the absolute covariant derivative $D\psi(x, e_i(x))$ of $\psi(x, e_i(x))$ possesses the same transformation behaviour under S(x) transformations as $\psi(x, e_i(x))$ itself with the transformation rule for the latter being expressed by eq. (3.39).

B. Second Definition of a Connexion

In this subsection we turn to the modern abstract mathematical definitions of a connexion in a principal fiber bundle over a manifold M_n. As a special case this will contain the linear connexion just discussed when one considers the connexion in the bundle of linear frames $L(M_n) = P(M_n, G = Gl(n,R))$ over the manifold. The more formal definition of a connexion in a principal fiber bundle will then in the last section of this chapter be extended to the so-called Cartan connexions. A special bundle over space-time possessing a Cartan connexion is used later in Chapter IV in an attempt to give strong interaction physics a basically differential geometric interpretation in terms of a gauge theory formulated on a fiber bundle with Cartan connexion characterized, as explained in the introduction, by an elementary length parameter which is associated with the range of the strong interaction forces.

We have defined in eqs. (3.1) two Lie groups of diffeomorphisms on a Lie group G, the group of left-translations and the group of right-translations of a point $g_1 \in G$ in the group G denoted by L_g and R_g, respectively. A left (or right-) invariant vector field X on the group G is defined by x)

$$dL_g\, X(g_1) = X(L_g g_1) = X(g g_1) \; , \qquad (3.111a)$$

$$dR_g\, X(g_1) = X(R_g g_1) = X(g_1 g) \; . \qquad (3.111b)$$

---

x) Footnote on p. 56.

The left-invariant (analogously, right-invariant) vector fields on G form a vector space of dimension n equal to the order of the group. One can define the Lie algebra $\mathfrak{g}$ of a group G as the space of left-invariant vector fields on G which is indowed with a product operation called the Lie product or Lie bracket. If X_a ; $a = 1,2 \dots n$, is a basis of the Lie algebra $\mathfrak{g}$ one has (compare eq. (2.41))

$$[X_a, X_b] = C^c_{ab} X_c \qquad (3.112)$$

with the structure constants C^c_{ab} obeying the relations (2.42) and (2.43) given in Section II.2. Stated differently the Lie algebra $\mathfrak{g}$ of a Lie group G ist, as a vector space, isomorphic to the tangent space $T_e(G)$ at the identity element e of the group together with the definition of a Lie bracket operation for the elements of this space as expressed by eq. (3.112). The mentioned isomorphism is provided by the translation in the group mapping X(g) into X(e).

A differential form ω defined on the group G is called left-invariant (analogously, right-invariant) if

$$d L^*_g \, \omega(g_1) = \omega(g g_1) \ . \qquad (3.113)$$

The space of all left-invariant one-forms on G is the dual space $\mathfrak{g}^*$ to the Lie algebra $\mathfrak{g}$. If ω^a; $a = 1,2 \dots n$, is a basis of one-forms spanning $\mathfrak{g}^*$ then the following formulae are true since with ω^a being left-invariant also the exterior derivative of ω^a is left-invariant, i.e. the two-forms $d\omega^a$ can be expanded in terms of $\omega^a \wedge \omega^b$ as

$$d\omega^c = -\tfrac{1}{2} C^c_{ab} \, \omega^a \wedge \omega^b \ . \qquad (3.14)$$

These formulae are called the equations of Maurer-Cartan. They are analogous to eqs. (3.112). The fact that the constants C^c_{ab} in eqs. (3.113) and (3.114) are the same follows from $dd\omega^c = 0$ and the Jacobi identities (2.43).

---

x) Here and in the following we denote quite generally by the symbol $d\varphi$ the mapping between the corresponding tangent spaces of the manifolds which are mapped into each other by the diffeomorphism φ .

We introduce one more concept before we turn to the general definition of an infinitesimal connexion in $P(M_n,G)$. The automorphism of G, called adj g, mapping ($g_1 \in G$)

$$g_1 \longrightarrow g g_1 g^{-1} \qquad \text{for every } g \in G$$

induces an automorphism of the Lie algebra $\mathfrak{g}$ which is also denoted by adj g. The representation provided by $g \longrightarrow$ adj g is called the adjoint representation of G in $\mathfrak{g}$.

A connexion in a principal fiber bundle $P(M_n,G)$ over M_n is defined by a linear mapping $\tilde{\tau}_u$ of the tangent space $T_x(M_n)$ to M_n at x into the tangent space $T_u(P)$ to $P(M,G)$ at $u \in \pi^{-1}(x)$ with the following properties:

1) $\tilde{\tau}_u(T_x(M_n)) = H_u$ is the horizontal subspace of $T_u(P)$ at $u \in F_x$.
2) $d\pi \circ \tilde{\tau}_u$ is the identity mapping .
3) If $u' = R_g u = ug$, with $g \in G$, then $\tilde{\tau}_{u'} = \tilde{\tau}_{ug} = dR_g \tilde{\tau}_u$.
4) $\tilde{\tau}_u$ is linear and depends differentiably on u.

The subspace $H_{u'}$ of horizontal vectors at $u' = R_g u = ug$ obtained from u by right translation x) in the fiber of the principal bundle by an element g of the structural group G is, according to property (3), given by

$$H_{u'} = H_{ug} = dR_g H_u \,. \tag{3.115}$$

Calling the tangent space to the <u>fiber</u> at $u \in F_x$ the vertical subspace of $T_u(P)$, i.e.

$$T_u(F_x(M_n)) = V_u \quad , \tag{3.116}$$

one obtains the following unique decomposition of any vector $X \in T_u(P)$ into horizontal and vertical components

$$X = X_v + X_h \tag{3.117}$$

---

x) We take the action of G in the fiber of $P(M_n,G)$ - being again G - to be <u>right-translation</u> in the group.

with $X_v \in V_u$ and $X_h \in H_u$ (compare the schematic drawing shown in Fig. 3) due to $d\pi(H_u) = T_x(M_n)$ and $d\pi(V_u) = 0$ implying that $T_u(P) = H_u \oplus V_u$.

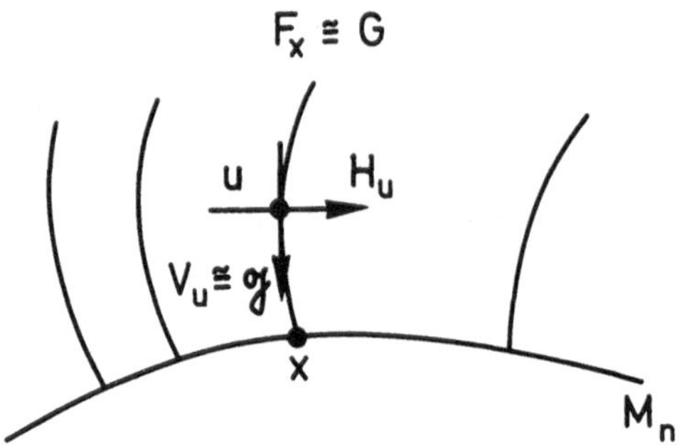

Fig. 3

According to eq. (3.117) a connexion in $P(M_n,G)$ can be defined by a differentiable mapping of $T_u(P)$ onto $T_u(F_x(M_n))$, the space tangent to the fiber at u, i.e. onto the vertical subspace of $T_u(P)$ which is isomorphic to the Lie algebra $\mathfrak{g}$ of the structural group G. This gives rise to a linear differential one-form ω with values in the Lie algebra of the structural group of the bundle called the form of the connexion on the principal fiber bundle $P(M_n,G)$. The one-form ω is characterized by the following properties

$$\omega(u\,dg) = g^{-1}dg \qquad (3.118a)$$

$$\omega(dR_g X) = g^{-1}\,\omega(X)\,g \qquad (3.118b)$$

where $u \in P(M_n,G)$, $\pi(u) = x$; $g \in G$, and $X \in T(P)$. In eq. (3.118a) dg is an element of $T_g(G)$ tangent to the group at g, and g^{-1} dg is the element of $T_e(G)$ tangent to the group at the identity element associated to dg, i.e. $g^{-1}dg \in \mathfrak{g}$ x). Eq. (3.118a) thus states that udg, being vertical, is mapped into the element of the Lie algebra corresponding to dg. Furthermore, eq. (3.118a) implies that

---

x) $g^{-1}(g+dg) = e+g^{-1}dg$.

$\omega(X_\cdot) = 0$ for X being horizontal. Denoting according to eq. (3.111a) the right-translation of the vector field X by $dR_g X$ eq. (3.118b) states that ω transforms according to the inverse adjoint representation, adj g^{-1}, of the group G in $\mathfrak{g}$.

To make contact with what has been said in subsection A one can now represent a connexion ω on $P(M_n,G)$ by a family of one-forms $\{\omega_{\mathcal{U}_i}\}$ where $\omega_{\mathcal{U}_i}$ is associated with a neighborhood U_i of a covering $\{\mathcal{U}_j\}$ for the base manifold M_n. Let $\sigma_i : M_n \rightarrow P(M_n,G)$ denote the local cross section over U_i in which $\omega_{\mathcal{U}_i}$ provides the connexion form in the gauge defined by σ_i. In the case of the bundle of linear frames $L(M_n) = P(M_n, G = Gl(n,R))$ we have called, in our previous discussion, the $\mathfrak{gl}(n,R)$- valued form, given in a certain neighborhood of M_n, by $\omega(x)$ possessing the transformation property (3.54'). There the matrix $a(x) \in Gl(n,R)$ describes the transformation in the intersection region of two local charts or the gauge transformation corresponding to a change of the local cross section from σ_i as given above to a new one σ'_i. In complete analogy one obtains here in the general case of an infinitesimal connexion in a principal fiber bundle $P(M_n,G)$ a family of $\mathfrak{g}$-valued one forms $\omega_{\mathcal{U}_i}$ on M_n, adapted to the covering $\{\mathcal{U}_j\}$ of M_n, having the transformation rule

$$\omega_{\mathcal{U}_i} = g_{ij}^{-1}\, \omega_{\mathcal{U}_j}\, g_{ij} + g_{ij}^{-1}\, dg_{ij} \tag{3.119}$$

where g_{ij} is an element of G depending differentiably on $x \in \mathcal{U}_i \cap \mathcal{U}_j$. Comparing eq. (3.119) with eq. (3.54') one notices a difference of the transformation involved i.e. a(x) corresponds to g_{ij}^{-1}. In the differential geometrical literature the form (3.119) is always given (compare also eqs. (3.118)) while our form (3.54') (see also eq. (3.110)), which we shall adhere to in the sequel, originated from the adopted convention expressed by the transformation formulae (3.53) with the lower index of the frame vectors e_i transforming with $a^{-1}(x)$ such that the contravariant vector components v^i transform with a(x) and not conversely.

We close this section by defining what is to be understood by a parallel displacement in a principal fiber bundle $P(M_n,G)$ with respect to a connexion ω. The definition of a connexion involved the mapping of a tangent vector of $T_x(M_n)$ at x into a tangent vector of $T_u(P)$ at $u \in F_x = \pi^{-1}(x)$. The connexion form ω was, moreover, characterized by the property that $\omega(X_h) = 0$ for X_h being a horizontal vector. We now call a vector field X(u) on $P(M_n,G)$ the <u>horizontal lift</u>, or simply the <u>lift</u>, of the vector field v(x) on M_n when i) X(u) is horizontal, i.e. $X(u) \in H_u$, for every $u \in F_x \subset P(M_n,G)$, and ii) X(u) is projected onto v(x) according to the formula

$$d\pi(X(u)) = v(\pi(u)) \quad . \tag{3.120}$$

This definition allows the introduction of the horizontal lift of a whole curve $C(t)$ on M_n given in local coordinates by $x^i = x^i(t)$; $i = 1,2 \ldots n$, with $a \leq t \leq b$, by lifting the tangent vectors $v^i(t) = \frac{dx^i(t)}{dt}$ of the curve $C(t)$ in the bundle space. One can thus define the horizontal lift of $C(t)$ as the piecewise differentiable curve $\gamma(t)$ in $P(M_n,G)$ the tangent vectors of which are all horizontal, i.e.

$$d\pi(X(t)) = v(t) \quad . \tag{3.121}$$

One writes accordingly

$$\pi(\gamma(t)) = C(t) \quad . \tag{3.122}$$

Denoting by x_o a point in the base space through which $C(t)$ passes (say for $t = 0$) one obtains in a unique way a lifted curve $\gamma(t)$ passing through $u_o \in \pi^{-1}(x_o)$. This can be seen as follows [16]: Since the bundle is locally trivial there is a curve $u(t)$ in $P(M_n,G)$ such that

$$\gamma(0) = u(0) = u_o \tag{3.123a}$$

and

$$\pi(\gamma(t)) = C(t) \quad . \tag{3.123b}$$

The lifted curve $\gamma(t)$ must be of the form

$$\gamma(t) = u(t)\, g(t) \tag{3.124}$$

in a general fiber bundle where $g(t)$ is a curve in the structural group G of the bundle such that $g(o) = e$ to satisfy eq. (3.123a). It is now the problem to determine a curve $g(t)$ in G in such a way that $\gamma(t)$ is horizontal. By differentiation with respect to the parameter t (denoted by a dot) it follows from eq. (3.124) that

$$\dot{\gamma}(t) = X(t) = \dot{u}(t)\, g(t) + u(t)\, \dot{g}(t) \quad . \tag{3.125}$$

Let now ω be the connexion form in $P(M_n,G)$. Then eqs. (3.118) imply that

$$\omega(X(t)) = g^{-1}(t)\, \omega(\dot{u}(t))\, g(t) + g^{-1}(t)\, \dot{g}(t) \tag{3.126}$$

with $g^{-1}(t)\dot{g}(t)$ being a curve in the Lie algebra $\mathfrak{g}$ of G. For $X(t)$ being horizontal one must have $\omega(X(t)) = 0$ which thus implies

$$\dot{g}(t)\, g^{-1}(t) = -\,\omega(\dot{u}(t)) \quad . \tag{3.127}$$

This equation can be integrated leading to a unique curve g(t) in the group G with g(o) = e . A unique curve g(t) in the structural group G of the bundle can thus be associated with a curve C(t) in the base space. This makes more precise the terminology used in Sect. II.1 in relation with the discussion of the Aharonov-Bohm experiment and the associated nonintegrable phase factor appearing in spinor electrodynamics.

One can now define a parallel displacement in the bundle $P(M_n,G)$ along the curve $\gamma(t)$ projecting onto the curve C(t) in the base space as follows: Consider the curve C(t) to join two arbitrary points x_o and x_1 of M_n. Let $u_o \in F_{x_o} = \pi^{-1}(x_o)$ be an arbitrary point in the fiber over x_o of $P(M_n,G)$, and let $\gamma(t)$ be the lift of C(t) starting from u_o [x]. Let the endpoint of $\gamma(t)$ in $P(M_n,G)$ corresponding to $x_1 \in M_n$ be u_1 with $\pi(u_1) = x_1$. By varying now the value u_o in the fiber $F_{x_o} = \pi^{-1}(x_o)$ one obtains a mapping, mediated by $\gamma(t)$, of the fiber over x_o onto the fiber $F_{x_1} = \pi^{-1}(x_1)$ over x_1 mapping u_o into u_1. This mapping is provided by an element g of the structural group uniquely associated with the path C(t) joining x_o and x_1 in M_n. The mapping $F_{x_o} \rightarrow F_{x_1}$ defines the parallel transport with respect to a given connexion ω (compare eq. (3.127)) of the fibers of $P(M_n,G)$ along a given curve in the base space of the bundle joining the points x_o and x_1. If the parallel transport is independent of the path joining any two given points in M_n the connexion is said to be <u>flat</u>. The necessary and sufficient condition for a connexion to be flat is the vanishing of the associated curvature two-form $\Omega = d\omega + \omega\wedge\omega$. (see eq. (3.72)).

III.3 Bundles with Cartan Connexions

In this section we turn to the discussion of a special type of fiber bundles which, on the one hand, marks the origin of the very concept of a fiber bundle in differential geometry in Cartan's work on generalized spaces possessing an affine, a projective, or a conformal connexion[26]. On the other hand, these Cartan-type bundles, as we would like to call them, seem to have attracted less interest in modern differential geometry which is reflected by the fact that they are not even mentioned in many of the text books on the subject. From a modern point of view these bundles

---

x) One could denote the lifted curve passing through $u_o \in F_{x_o}$ by $\gamma_{u_o}(t)$. Then for any other point $u = u_o g$ of F_{x_o} one has

$$\gamma_u(t) = R_g\, \gamma_{u_o(t)} = \gamma_{u_o g}(t) .$$

were discussed by Ehresmann[27] and by Kobayashi[19]. Ehresmann has introduced the name bundles with Cartan connexion for them.

The interest we have for these Cartan-type bundles in relation to physics is that it appears that the geometry in such a bundle provides a framework for a gauge description of strong interaction phenomena expressible by means of differential geometrical concepts and techniques as outlined in the introduction. One avoids thereby the puzzles associated with the assumption of real constituents existing inside hadrons determining the hadronic interaction in terms of the point-like physics of these conjectured subunits of hadrons. Admittedly the constituent concept was extremely fruitful in physics and has led us from molecules to atoms, from atoms to the electrons and the atomic nucleus, and, finally, from the atomic nucleus to the individual nucleons i.e. to protons and neutrons. The question is whether such a chain goes on further and further or whether it ends somewhere (for exemple at the level of hadrons) and nature is, indeed, able to manufacture objects possessing extension, i.e. having space-filling properties, already at a fundamental level without requiring the existence of subunits or constituents with point-like interaction for building them. I would dare saying that the key problem in todays physics is the setting up a convincing theory for <u>extended</u> hadrons, possessing a mass and spin spectrum, in terms of a theoretical framework avoiding such concepts as that of point-like constituents which, experimentally, seem never to appear isolated in nature and thus are not constituents in the true sense of this word.

The outstanding feature of a bundle with Cartan connexion which make these generalized spaces so interesting as a possible geometrical stratum on which hadron physics could manifest itself is the fact that the fiber over x in a Cartan-type bundle is <u>tangent</u> to the base space at x allowing thus the coordinates in F_x to play the role of generalized relative coordinates, i.e. coordinates relative with respect to the point of contact x of base space and fiber. In the next chapter we shall investigate a special bundle of Cartan type over space-time as base space and use it for a differential geometric description of hadronic interactions. In this section we give the necessary mathematical pleniminaries.

We first define the concept of the soldering [x)] for a bundle of Cartan type[27,19]. A fiber bundle [xx)] $E(M_n,F,G,P)$ over M_n associated to the principal fiber bundle $P(M_n,G)$ is called <u>soldered</u> to M_n if the following conditions are satisfied:

---

x) French: soudure

xx) For brevity we leave out the projection π and the family of homeomorphisms ϕ in the specification of the bundle.

1) The group G acts transitively on F, i.e. F is the homogenous space G/G' where G' is the stability subgroup of G leaving the point O of F fixed.
2) dim F = dim M_n = n.
3) The bundle $E(M_n,F,G,P)$ admits a cross section which will be identified with M_n.
4) If $T'(M_n)$ is the space of all tangent vectors to F_x at $x \in M_n$ for all x and $T(M_n)$ is the tangent bundle over M_n then one can identify $T'(M_n)$ and $T(M_n)$ by an isomorphism.

The property (4) states that the fiber over x is tangent to the base space at x for every $x \in M_n$. $T'(M_n)$ is a fiber bundle over M_n with fiber R_n and structural group $Gl(n,R)'$ which can be identified with the space $(P' \times R_n)/G'$ being a fiber bundle associated with $P'(M_n,G')$ x). Here the principal bundle $P'(M_n,G')$ is a subspace of $P(M_n,G)$ obtained by restricting the homeomorphisms $F \to F_x$ of $P(M_n,G)$ in such a way that the point $O \in F$ is always mapped into x, i.e. into the point of contact of fiber and base space at $x \in M_n$. To the principal fiber bundle $P'(M_n,G')$ with structural group G' is associated in the usual way a bundle $E(M_n,F,G',P')$ the existence of which was first shown by Ehresmann to define the soldering in $E(M_n,F,G,P)$.

Let now ω denote the form of a Cartan connexion in $P(M_n,G)$ (and thereby in $E(M_n,F=G/G',G,P)$), and let $\bar{\omega}$ be the restriction of ω to the bundle $P'(M_n,G')$ (and thereby to $E(M_n,F=G/G',G',P')$. Let, furthermore, $\mathfrak{g}'$ denote the Lie algebra of G'. Then $\bar{\omega}$ is a $\mathfrak{g}$-valued linear differential one-form satisfying the following conditions:

$$\bar{\omega}(u'dg') = g'^{-1}dg' \tag{3.128a}$$

x) There exists a mapping of G' (viewed as a transformation group in F leaving $O \in F$ fixed) into $Gl(n,R)'$ (viewed as a transformation group in $T_o(F)$) which can also be used to associate $E(M_n,F,G',P')$ (see below) to $T'(M_n)$, i.e. the mentioned isomorphism consists in identifying the group G' with $Gl(n,R)'$. The group $Gl(n,R)'$ operating in $T_o(F)$, and the group $Gl(n,R)$, operating in $T(M_n)$, are conceptually two different gauge groups. The former belonging, as a subgroup of G, to the fiber of the Cartan-type bundle, the latter belonging to the fiber of the linear frame bundle over M_n.

$$\bar{\omega}(dR_{g'}X') = g'^{-1}\,\bar{\omega}(X')\,g' \tag{3.128b}$$

If $\bar{\omega}(X') = 0$ then X' is the zero vector. (3.128c)

Here $u' \in P'(M_n,G')$, $\pi(u') = x$ [with π' denoting the projection in $P'(M_n,G')$], $g' \in G'$, $dg' \in T_{g'}(G')$, and $X' \in T(P')$. Eqs. (3.128a) and 3.128b) state that $\bar{\omega}$ defines a connexion in $P'(M_n,G')$, and eq. (3.128c) implies that any horizontal vector in $P'(M,G')$ is the zero vector. This last mentioned property is a consequence of the soldering of $E(M_n,F,G',P')$ to M_n.

Starting from eqs. (3.128a-c) it is now easy to specify uniquely the form of a Cartan connexion in $E(M_n,F,G,P)$ associated to $P(M_n,G)$ by the following $\mathfrak{g}$-valued differential form

$$\omega(X) = g^{-1}\,\bar{\omega}(X')\,g + g^{-1}dg \tag{3.129}$$

where $g \in G$, $dg \in T_g(G)$, and $X = dR_g X'$ with $X' \in T(P')$ and $X \in T(P)$. It is seen from eq. (3.129) that the restriction of ω to $P'(M_n,G')$ is $\bar{\omega}$.

Let us now decompose the Lie algebra $\mathfrak{g}$ of G into the subalgebra $\mathfrak{g}'$ and a vector subspace $\mathfrak{s}$, i.e.

$$\mathfrak{g} = \mathfrak{g}' \oplus \mathfrak{s} \ . \tag{3.130}$$

Then the tangent space $T_o(F)$ at $o \in F$ is isomorphic to $\mathfrak{s}$. If, furthermore

$$[\mathfrak{g}', \mathfrak{s}] \subseteq \mathfrak{s} \tag{3.131}$$

as will be the case for the bundle used in Chapter IV for a description of hadronic physics, then $T_o(F)$ can be identified with $\mathfrak{s}$ and there exists a linear $\mathfrak{s}$-valued differential form on $P'(M_n,G')$, called the <u>form of soldering</u> with the properties [x)]

---

x) If eq. (3.131) does not hold true one can only say that $g' \in G'$ induces a linear transformation $L_{g'}$ of $T_o(F)$ which does not correspond to the adjoint representation of G' in $T_o(F)$. In this case eq. (3.132b) reads $\theta(dR_{g'}X') = L^{-1}_{g'}\,\theta(X')$ with $\theta(X')$ being a $T_o(F)$-valued form.

$$\theta(X') = 0 \quad \text{for } X' \in T(P') \text{ if and only if } d\pi'(X') = 0 \tag{3.132a}$$

$$\theta(dR_{g'}X') = g'^{-1}\,\theta(X')\,g' \quad \text{for all } X' \in T(P') \text{ and } g' \in G' \tag{3.132b}$$

Eq. (3.132a) states that θ vanishes for all vertical vectors on $P'(M_n,G')$.

It can be shown[19] that if the bundle $E(M_n, F=G/G',G,P)$ satisfies the conditions (1), (2), and (3), and if the Lie algebra $\mathfrak{g}$ of G and the Lie algebra $\mathfrak{g}'$ of the stability subgroup G' of G decomposes according to eq. (3.130), obeying eq. (3.131), then the bundle $E(M_n, F=G/G',G,P)$ is soldered to M_n (according to property (4)) if and only if there exists a $\mathfrak{d}$-valued linear differential form θ satisfying eqs. (3.132a) and (3.132b).

The interest in the form θ lies in the fact that, if ω' is a $\mathfrak{g}'$-valued one-form defining a connexion in $P'(M_n,G')$,

$$\bar{\omega} = \omega' + \theta \tag{3.133}$$

is a $\mathfrak{g}$-valued one-form defining in a one-to-one way the restriction to $P'(M_n,G')$ of a Cartan connexion ω in $P(M_n,G)$ (and, correspondingly, in the associated bundle $E(M_n,F=G/G',G,P)$ soldered to M_n). The proof of this statement, i.e. that $\bar{\omega}$, as defined by eq. (3.133), satisfies eqs. (3.128a-c) follows directly from the properties (3.132a) and (3.132b) and the definition of the connexion ω' in $P'(M_n,G')$ (see eqs. (3.118)).

To conclude our review of Cartan bundles and their soldering property we finally write down Cartan's structural equations for the space $E(M_n,F=G/G',G,P)$

$$d\bar{\omega} + \tfrac{1}{2}[\bar{\omega},\bar{\omega}] = \bar{\Omega}. \tag{3.134}$$

$\bar{\Omega}$ is here the curvature two-form of the connexion $\bar{\omega}$ and $[,]$ denotes the exterior product of forms with values in a Lie algebra x). Using in eq. (3.134) the decompo-

---

x) Let α_ς ; $\varsigma = 1,2 \ldots r$ be a basis of a Lie algebra $\mathfrak{g}$, and let A and B be $\mathfrak{g}$-valued differential forms $A = a^\varsigma \alpha_\varsigma$; $B = b^\varsigma \alpha_\varsigma$, where a^ς is a p-form and b^ς is a q-form, then $[A,B]$ is a (p+q)-form defined by

$$[A,B] = a^\varsigma \wedge b^{\varsigma'} \, [\alpha_\varsigma, \alpha_{\varsigma'}]$$

where the bracket on the right-hand side of this equation is the Lie bracket defined in $\mathfrak{g}$. It follows from the antisymmetry of the Lie bracket that

$$[A,B] = A\wedge B - (-1)^{qp} B\wedge A = (-1)^{qp+1}[B,A].$$

Thus one has, for example,

$$\tfrac{1}{2}[\omega,\omega] = \omega\wedge\omega.$$

sition (3.133) and separating both sides of the equation into a $\mathfrak{g}'$-valued and a $\mathfrak{s}$-valued part, remembering eq. (3.131), and putting, moreover,

$$\bar{\Omega} = \bar{\Omega}_{\mathfrak{g}'} + \bar{\Omega}_{\mathfrak{s}} \tag{3.135}$$

and, similarly,

$$[\theta,\theta] = [\theta,\theta]_{\mathfrak{g}'} + [\theta,\theta]_{\mathfrak{s}} \tag{3.136}$$

one obtains the eqs. (3.138) and (3.139) below by making, furthermore, use of the structural equations (3.72) written in the form

$$d\omega' + \tfrac{1}{2}[\omega',\omega'] = \Omega' \tag{3.137}$$

as defined by the connexion ω' in $P'(M_n,G')$; i.e.:

$$\Omega' = \bar{\Omega}_{\mathfrak{g}'} - \tfrac{1}{2}[\theta,\theta]_{\mathfrak{g}'}\,, \tag{3.138}$$

$$d\theta + \tfrac{1}{2}[\omega',\theta] + \tfrac{1}{2}[\theta,\omega'] = D'\theta = \tau\,. \tag{3.139}$$

Here

$$\tau = \bar{\Omega}_{\mathfrak{s}} - \tfrac{1}{2}[\theta,\theta]_{\mathfrak{s}} \tag{3.140}$$

denotes the torsion form of the connexion $\bar{\omega}$ and $D'\theta = d\theta + [\omega',\theta]$ represents the exterior covariant derivative of the form θ with respect to the connexion ω'.

If now the space F=G/G' satisfies, furthermore, the condition[x)]

$$[\mathfrak{s},\mathfrak{s}] \subseteq \mathfrak{g}' \tag{3.141}$$

as will be the case for the physically interesting example of a Cartan-type bundle investigated in Chapter IV, the eqs. (3.138) and (3.140) reduce to the form

---

x) Spaces satisfying eqs. (3.130), (3.131) and (3.141) together with $[\mathfrak{g}',\mathfrak{g}'] \subseteq \mathfrak{g}'$, the latter expressing the fact that $\mathfrak{g}'$ is a subalgebra of $\mathfrak{g}$, are called globally symmetric Riemannian spaces (compare refs. 28 and 29).

$$\bar{\Omega}_{g'} = \Omega' + \tfrac{1}{2}[\theta,\theta] \tag{3.142}$$

$$\bar{\Omega}_{\mathfrak{a}} = \tau . \tag{3.143}$$

If, one the other hand,

$$[\mathfrak{a},\mathfrak{a}] = 0 \tag{3.144}$$

which is a condition valid for the affine tangent bundle $T_A(M_n)$ with fiber $F=A(n,R)/Gl(n,R)=R_n$ where $A(n,R)$ denotes the group of affine transformations in the R_n, composed of the general linear transformations of the basis in R_n together with the translations of the origin of the frame, one obtains

$$\bar{\Omega}_{g'} = \Omega' , \tag{3.145}$$

$$\bar{\Omega}_{\mathfrak{a}} = \tau . \tag{3.146}$$

For a general discussion of an Inönü-Wigner contraction process of the structural group G with respect to the stability subgroup G' in the fiber of a general Cartan-type fiber bundle and the resulting relation between $E(M_n,F=G/G',G,P)$ and the affine tangent bundle $T_A(M_n)$ associated to the bundle of affine frames $P(M_n,G=A(n,R))$ over M_n we refer to the literature[30].

IV. GAUGE DESCRIPTION OF STRONG INTERACTIONS BASED ON A FIBER BUNDLE OF CARTAN TYPE

This chapter is devoted to an application of the differential geometrical formulation developed in the previous sections to the physics of hadrons in an attempt to establish a gauge theory for the strong interactions. As described in the introduction we shall choose as the underlying geometric substratum a higher dimensional space constructed over a Riemannian space-time, V_4, of general relativity (in the presence of gravitation), or over Minkowski space-time, M_4, of special relativity (when long range gravitational effects can be considered negligible) which contains degrees of freedom capable of representing strong interaction phenomena of extended elementary structures. Stated differently, we shall assume that the physics of strong interactions between the individual constituents of matter in the form of extended hadrons manifest itself on a generalized space having the structure of a fiber bundle of Cartan type constructed or space-time. In most of the following discussion we shall, moreover, assume that gravitational effects, although in principle contained in the formalism, can be disregarded in focussing the attention on the short range proporties of matter immersed in the fiber bundle geometry with the fibers being associated here with effects related to strong interaction phenomena. We thus regard in most of our discussion the base manifold of the Cartan-bundle — the latter specifying, as said, the geometry in a world in which hadron physics is conceived to evolve — to be flat Minkowski space-time M_4. Ultimately, however, the capacity of matter to generate long range gravitational fields, i.e. to produce a Riemannian curvature field in the space-time base manifold of the bundle, has to be included in the theory when a full description is intended to be given which is valid also for large scale phenomena of matter present in the geometry.

The bundle relevant for a gauge description of the strong interactions is a fiber bundle over space-time with Cartan connexion possessing as fiber over each space-time point x a local four-dimensional Riemannian space, $V_4'(x)$, of constant curvature, with radius of curvature $R \approx 10^{-13}$ cm, possessing the de Sitter group SO(4,1) as structural group. For short we call this bundle the de Sitter bundle over space-time. For the discussion of the properties of hadronic matter at small distances we shall introduce a generalized matter or wave field, $\psi(x,\xi)$, which can be defined as a cross section on a bundle associated with the de Sitter bundle in the same way as vector or spinor fields in general relativity can be considered as cross sections on bundles. After defining in Sect. IV.1 the de Sitter bundle over space-time in terms of the notation established in Chapter III we turn, in Sect. IV.2, to a mathematical formulation of the dualism between hadronic matter, on the one side, and the underlying fiber bundle geometry, on the other side, which was described qualitatively in the introduction, and called there strong fiber dynamics (SFD), leading to the basic current curvature equations. In Sect. IV.3

we, finally, give a combined description of electromagnetic and strong interactions in terms of a U(1) and USp(2,2) (de Sitter) gauge invariant Lagrangian formalism for charged hadronic spinor matter fields.

IV.1 The de Sitter Bundle over Space-Time

The fiber bundle

$$E\,(V_4,\, F = SO(4,1)/SO(3,1),\, G = SO(4,1),\, P) \tag{4.1}$$

associated to the prinicipal bundle $P = P(V_4, SO(4,1))$ is a bundle of Cartan type possessing as fiber the noncompact coset space SO(4,1)/SO(3,1) isomorphic to a de Sitter space, V_4', on which the group SO(4,1) acts as a group of motion. The space V_4' can be embedded into a five-dimensional pseudo-Euklidian space $R_{4,1}$ and represents a hypersurface there, the so-called de Sitter (hyper)-hyperboloid, given by [x)]

$$\xi^a \xi_a = \xi^a \xi^b \eta_{ab} = -R^2 \tag{4.2}$$

where $\xi^a = (\xi^0, \xi^1, \xi^2, \xi^3, \xi^5)$ denote the coordinates in $R_{4,1}$ (and by eq. (4.2) on V_4'), and η_{ab} = diag (1,-1,-1,-1,-1) represents the metric tensor in $R_{4,1}$. The constant R in eq. (4.2) denotes the radius of curvature of the hyperbolic space V_4' with V_4' representing the standard fiber of the de Sitter bundle. In the same way as one says that the hyperboloid (4,2) is a model for the de Sitter space one can say that the following union of local de Sitter spaces $V_4'(x)$, being diffeomorphic images of the standard V_4', taken for all points x of space-time provides a model for the de Sitter bundle (4,1) [31)]

$$T^R(V_4) = \bigcup_{x \in V_4} V_4'(x)\ . \tag{4.3}$$

The soldering property is transparently expressed through the notation (4,1) (together with the discussion given in Sect. III.3) with the Lorentz group G' = SO(3,1) being the stability subgroup of SO(4,1) in F corresponding to the point of contact $O = \overset{\circ}{\xi}$ identified with $x \in V_4$ in the isomorphism relating the tangent planes

---

x) Upper and lower indices a,b,c,... are, in this chapter, to be summed over 0,1,2,3,5, while upper and lower indices i,j,k,... are only summed over 0,1,2,3.

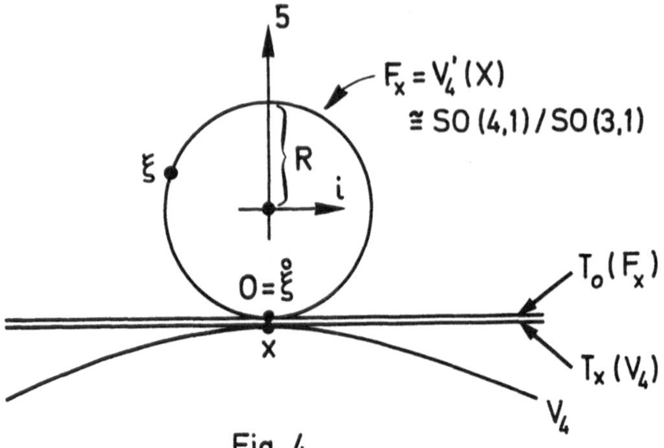

Fig. 4

$T_o(F_x)$ and $T_x(V_4)$ [see Fig. 4 where we have still drawn separately these two tangent planes which are to be identified by the isomorphism defining the soldering of the bundle]. On the other hand eq. (4.3) provides a model for the bundle (4,1) characterized by a definite value for R. It is this model, with R having a fixed definite value (independently of x) of the order of one Fermi, that we intend to use as a geometrical framework on which hadronic matter fields are to be defined. Thus, as stated in the introduction, our underlying space on which we want to formulate a generalized wave motion, meant to describe the translational and internal motions of extended hadrons, is the eight-dimensional fiber space (4,3) with the velocity of light characterizing the local Minkowski geometry in the base space and the curvature radius R, playing the role of an elementary length parameter associated with strong interactions, characterizing globally the fiber of the bundle.

There is still one more remark to be made to define the de Sitter bundle uniquely. This is connected with the possibility to choose orientations in the base space and in the fiber. As mentioned before, we shall assume space-time to be orientable. Since the full Lorentz group (including space and time reflexions) as well as the full de Sitter group (including "internal space" and "internal time" reflexions) are both four-fold connected one can solder the de Sitter bundle in 16 possible ways, labelled [x)]

---

x) In the first bracket, referring to the base space, T = +(-) means positive (negative) time orientation, and S = L(R) means left (right) handedness of the spacial vectors of the local frames. Analogously, the second bracket determines the orientation characterizing the fiber (V_4').

$$[T,S]\,;\,[\tau,\sigma] \qquad ;\begin{cases} T=\pm\,,\ S=L,R \\ \tau=\pm\,,\ \sigma=\ell,r \end{cases} , \tag{4.4}$$

which are obtained from what one could call the standard orientation, given by $[+,R]$; $[+,r]$, in terms of 16 symmetry operations defined globally on the bundle x). As long as we do not speak of discrete symmetry transformations applied to quantities defined on the bundle space we shall always assume that the orientation chosen is the standard one. However, whatever orientation one chooses among the 16 possibilities given by (4.4) the structural groups of the Lorentz frame bundle over V_4,

$$L(V_4) = P(V_4\,, G = O(3,1)^{++}) \tag{4.5}$$

and the de Sitter frame bundle over V_4,

$$L^R(V_4) = P(V_4\,, G = O(4,1)^{++}) \tag{4.6}$$

are in the notation of Sect. III.1, $O(3,1)^{++}$ and $O(4,1)^{++}$, respectively, as indicated by the right-hand sides of eqs. (4.5) and (4.6).

Before we relate the bundles $T^R(V_4)$ and $L^R(V_4)$ in the limit $R \to \infty$ to the affine tangent bundle $T_A(V_4)$ and the bundle of affine frames, respectively, where $T_A(V_4)$ has been defined at the end of the previous section and $L_A(V_4) = P(V_4, A(n,R))$, we write down the Lie algebra of the group SO(4,1) and introduce the decomposition (3.130). Denoting by

$$M_{ab} = -M_{ba} \qquad ; a,b = 0,1,2,3,5 \tag{4.7}$$

the ten generators of SO(4,1), the Lie algebra $\mathfrak{so}(4,1)$ is defined by the commutation relations

$$i[M_{ab}, M_{cd}] = \eta_{ac} M_{bd} + \eta_{bd} M_{ac} - \eta_{ad} M_{bc} - \eta_{bc} M_{ad} . \tag{4.8}$$

Introducing

$$\Pi_i = \frac{1}{R} M_{5i} \qquad ; \ i = 0,1,2,3 \tag{4.9}$$

---

x) Compare App. A of ref. 32 for a more detailed discussion.

eqs (4.8) can be written, with η_{ik} = diag (1,-1,-1,-1), as

$$i[M_{ij}, M_{kl}] = \eta_{ik} M_{jl} + \eta_{jl} M_{ik} - \eta_{il} M_{jk} - \eta_{jk} M_{il} \tag{4.10a}$$

$$i[\Pi_i, M_{jk}] = \eta_{ik} \Pi_j - \eta_{ij} \Pi_k \tag{4.10b}$$

$$i[\Pi_i, \Pi_j] = - \frac{1}{R^2} M_{ij} \tag{4.10c}$$

corresponding to the formulae $[\mathfrak{g}', \mathfrak{g}'] \subseteq \mathfrak{g}'$, $[\mathfrak{g}', \mathfrak{a}] \subseteq \mathfrak{a}$ and $[\mathfrak{a}, \mathfrak{a}] \subseteq \mathfrak{g}'$ of Sect. III.3, respectively. Here $\mathfrak{g}'$ is spanned by the M_{ij}; i,j = 0,1,2,3, generating the Lorentz subalgebra (4.10a), and $\mathfrak{a}$ is spanned by the Π_i ; i = 0,1,2,3, defined in eq. (4.9). It is now easy to carry out an Inönü-Wigner contraction [x)] of the de Sitter algebra corresponding to the singular limit $R \to \infty$ in which the algebra (4.10) goes over into the algebra of the Poincaré group ISO(3,1) given by

$$i[M_{ij}, M_{kl}] = \eta_{ik} M_{jl} + \eta_{jl} M_{ik} - \eta_{il} M_{jk} - \eta_{jk} M_{il} \tag{4.11a}$$

$$i[P_i, M_{jk}] = \eta_{ik} P_j - \eta_{ij} P_k \tag{4.11b}$$

$$[P_i, P_j] = 0 \tag{4.11c}$$

with the Π_i of eqs (4.9) going over in the contraction limit into the translation operators P_i. This discussion shows that both ten parameter groups, SO(4,1) and ISO(3,1), are related by

$$SO(4,1) \xrightarrow{R \to \infty} ISO(3,1) \quad . \tag{4.12}$$

Correspondingly, one obtains for the de Sitter bundle (4.3) and the associated de Sitter frame bundle (4.6) the contraction relations [30)]

---

x) Compare, for example, ref. 28

$$T^R(V_4) \xrightarrow{R \to \infty} T_A(V_4) \;, \tag{4.13}$$

$$L^R(V_4) \xrightarrow{R \to \infty} L_A(V_4) \;. \tag{4.14}$$

I.e. in the contraction limit one obtains the affine tangent bundle $T_A(V_4)$ and the affine frame bundle over space-time. The former is a special Cartan-type bundle associated to the latter with both bundles possessing as structural group the Poincaré group ISO(3,1) being the semidirect product of the Lorentz group [x)] and the Minkowski space (i.e. the translations). One can thus characterize physically the new degrees of freedom possible on the de Sitter bundle $T^R(V_4)$ by saying that the generators Π_i [xx)] (which in the limit $R \to \infty$ correspond to the abelian translation operators P_i acting in the flat fiber of the affine tangent bundle) are members of a set of generators of a ten-parameter semisimple group (the de Sitter group SO(4,1)) which allow the possibility of generalized "translational" gauge degrees of freedom to be considered in the formalism. They are no true translational degrees of freedom, of course. Due to the smallness of R they are confined to a neighbourhood of the point $x \in V_4$ determined by R when applied to F_x, the fiber over x, as is transparent from the schematic drawing of the de Sitter bundle given in Fig. 4. The last statement can be made more precise when one stereographically projects the local fiber F_x onto the common tangent space at the point x (compare ref. 31). Then the rotation given by M_{5i}, i.e. the de Sitter boost (see eq. (4.9)), indeed corresponds to a translation in the projection coordinates.

Since we want to introduce spinor fields on the de Sitter bundle constructed over space-time we now turn to the definition of a spinor structure, or spinor bundle, associated with the de Sitter frame bundle (4.6). To do this we first introduce the covering group $\overline{SO(4,1)}$ = USp(2,2) of SO(4,1) and discuss the basic four component de Sitter spinors $\phi(\xi, e_a(\xi)) = (\phi^{A'}(\xi, e_a(\xi))\;;\; A' = 1,2,3,4)$ transforming under USp(2,2). The spinor $\phi(\xi, e_a(\xi))$ is referred to a local frame at the point $\xi \in V_4'$ given by

$$e_a(\xi)\;;\; a = 0,1,2,3,5\;, \text{ obeying } \xi^a e_a(\xi) = 0\,. \tag{4.15}$$

---

x) More exactly, the group $O(3,1)^{++}$ of special orthochronous Lorentz transformations.

xx) The Π_i generate special de Sitter transformations called previously [31)] the de Sitter boosts.

The de Sitter transformations act as global hyperbolic rotations in the embedding space $R_{4,1}$ leaving the hyperboloid V_4', expressed by eq. (4.2), invariant. A spinor $\phi(\xi, e_a(\xi))$ attached to a point and referred to a system of axes at this point will suffer a global transformation on V_4' under de Sitter transformations being thereby moved into another point $\xi' \in V_4'$ and referred to a transformed system of axes. Before we formulate this let us briefly discuss the group USp(2,2) and its relation to SO(4,1).

The group USp(2,2) is the intersection of the Lie groups U(2,2) and Sp(4,C) [x)], i.e.

$$USp(2,2) = U(2,2) \cap Sp(4,C) \quad . \tag{4.16}$$

If $A \in SO(4,1)$ and $\hat{S}(A) \in USp(2,2)$, with $\hat{S}(A)$ denoting a transformation of the covering group associated to A, then the group homomorphism $USp(2,2) \rightarrow SO(4,1)$ is expressed by the formula analogous to eq. (3.36):

$$\gamma^a [A^{-1}]_a^{\ b} = \hat{S}(A)\, \gamma^b\, \hat{S}^{-1}(A) \tag{4.17}$$

where $\gamma^a = (\gamma^i, \gamma^5)$ with $\gamma^5 = \gamma^0\gamma^1\gamma^2\gamma^3$ are the five anticommuting Dirac γ-matrices obeying (compare eq. (2.7))

$$\{\gamma^a, \gamma^b\} = 2\eta^{ab} \cdot 1 \quad , \tag{4.18}$$

$$\gamma^{a\dagger} = \gamma^0 \gamma^a \gamma^0 \quad . \tag{4.19}$$

The four component de Sitter spinors $\phi(\xi, e_a(\xi))$ and $\bar{\phi}(\xi, e_a(\xi))$ transform under de Sitter transformations as

$$\phi'(\xi', e_a'(\xi')) = \hat{S}(A)\, \phi(\xi, e_a(\xi)) \tag{4.20a}$$

$$\bar{\phi}'(\xi', e_a'(\xi')) = \bar{\phi}(\xi, e_a(\xi))\, \hat{S}^{-1}(A) \tag{4.20b}$$

---

x) USp(2,2) is isomorphic to a subgroup of Gl(2,Q), the general linear group of 2 x 2 matrices over the field of quaternions.[33)] with the adjoint spinor being defined by

$$\bar{\phi}(\xi, e_a(\xi)) = \phi^+(\xi, e_a(\xi))\, \gamma^0 \tag{4.21}$$

and $\hat{S}^{-1}(x)$ being given by the relation

$$\hat{S}^{-1}(A) = \gamma^0\, S^+(A)\, \gamma^0 \tag{4.22}$$

expressing the non-unitarity of the finite dimensional x) spinor representation of SO(4,1) provided by the matrices $\hat{S}(A)$. In eqs. (4.20) ξ' and $e'_a(\xi')$ are given by

$$\xi'^a = [A]^a_b\, \xi^b \quad , \tag{4.23a}$$

$$e'_a(\xi') = [A^{-1}]^b_a\, e_b(\xi) \tag{4.23b}$$

where A leaves the de Sitter hyperboloid (4.2) invariant, i.e.

$$\xi'^a \xi'_a = \xi^a \xi_a = -R^2 \tag{4.24}$$

implying that

$$A^a_c\, A^b_d\, \eta_{ab} = \eta_{cd} \quad . \tag{4.25}$$

We, finally, quote the matrix elements of the generators $R^{ab}=-R^{ba}$ of SO(4,1) in the basic 5x5 representation:

$$[R^{ab}]^c_d = -i\left(\delta^a_d \delta^b_e - \delta^b_d \delta^a_e\right)\eta^{ec} \quad . \tag{4.26}$$

In terms of the Dirac γ-matrices γ^a the 4x4 spinor representation of the algebra **so**(4,1) is given by the matrices

$$M^{ab} = \frac{i}{4}[\gamma^a, \gamma^b] \; ; \; a,b = 0,1,2,3,5. \tag{4.27}$$

---

x) I.e. four-dimensional.

In complete analogy to the definition of a spinor structure of Dirac four component type in general relativity (compare eq. (3.35)) we now define a de Sitter spinor structure S^R over space-time by the following spinor bundle

$$S^R(V_4, F=C_4, G=USp(2,2), L^R) \qquad (4.28)$$

associated to $L^R(V_4)$ and, of course, also associated to $T^R(V_4)$. The fiber C_4 of S^R is a complex linear space, with elements $\phi(\xi, e_a(\xi))$, on which the group USp(2,2) acts in the sense of eqs. (4.20) as a transformation group. Expressed in words S^R represents the totality of de Sitter spinor fields referred to all possible systems of de Sitter frames (described by $L^R(V_4)$) which may be introduced in a differentiable fashion on the Cartan bundle $T^R(V_4)$. Choosing a local cross section in $L^R(V_4)$ consists, in the terminology introduced previously, in choosing a local de Sitter gauge on $T^R(V_4)$, namely, choosing in a neighbourhood U of V_4 a certain de Sitter frame in each contact point $O \in F_x$, identified with $x \in V_4$, and boosting it over the local fiber F_x by the action of the group $O(4,1)^{++}$ x). We write a system of base vectors on F_x in the same way as in eq. (4.15) suppressing the dependence on x.

A representative of a spinor field defined on $T^R(V_4)$ (possessing scalar Lorentz character xx)) can now be given as a cross section in S^R, i.e.

$$\Phi(x,\xi) = \phi(x;\xi, e_a(\xi)) \quad . \qquad (4.29)$$

Performing a de Sitter gauge transformation, corresponding to an x-dependent rotation (4.23) of the local de Sitter space as a whole, one obtains in analogy to eq. (3.39) the transformation rules for the representatives $\phi(x;\xi,e_a(\xi))$ and $\Phi(x;\xi,e_a(\xi))$:

$$\phi'(x;\xi',e'_a(\xi')) = \hat{S}(x)\, \phi(x;\xi,e_a(\xi)) \qquad (4.30a)$$

---

x) $L^R(V_4)$ describes in each space-time point x a ten-parameter family of frames: I.e. a six-parameter family of frames in each point ξ of $F_x = V_4'(x)$ with $V_4'(x)$ being itself a four-parametrical hypersurface.

xx) Since we assume ϕ to behave as a Lorentz scalar under local Lorentz transformations in the base space [or, more exactly, under transformations on the Lorentz frame bundle over V_4] we do not mention in eq. (4.29) the local Lorentz frame $e_i(x)$ in addition to the variable x.

$$\bar{\Phi}'(x;\xi',e'_a(\xi)) = \bar{\Phi}(x;\xi,e_a(\xi))\,\hat{S}^{-1}(x) \tag{4.30b}$$

with $\hat{S}(x) = \hat{S}(A(x))$ and eqs. (4.23) for the transformed point ξ' and the transformed frame $e'_a(\xi')$, and the correspondence between $\hat{S}(x)$ and $\hat{A}(x)$ as determined by eq. (4.17).

In our later discussion we shall make use of bispinor quantities, $\psi^{AA'}$, with A=1,2,3,4 being a Dirac spinor index, transforming with S(x) like in eq. (3.39), and with A'=1,2,3,4 being a de Sitter spinor index transforming with $\hat{S}(x)$ as shown in eq. (4.30a). The Dirac-de Sitter bispinor quantity and its adjoint will thus be written, in a certain gauge, as

$$\psi(x,e_i(x);\xi,e_a(\xi)) \tag{4.31a}$$

$$\bar{\bar{\psi}}(x,e_i(x);\xi,e_a(\xi)) \tag{4.31b}$$

or simply as $\psi(x,\xi)$ and $\bar{\bar{\psi}}(x,\xi)$ with

$$\bar{\bar{\psi}}(x,\xi) = \psi^{\dagger}(x,\xi)\,\gamma^0\otimes\gamma^0 \tag{4.32}$$

where an evident direct product notation for the γ-matrices associated with the Dirac (space-time) and de Sitter (fiber) part of $\psi(x,\xi)$ is used. As indicated above there are two different gauge groups at work now transforming a bispinor quantity (4.31) defined on the de Sitter bundle: a) the Lorentz gauge transformations, S(x), affecting the unprimed spinor index of $\psi(x,\xi)$ (compare eq. (3.39)), and b) the de Sitter gauge transformations, $\hat{S}(x)$, affecting the primed spinor index of $\psi(x,\xi)$. Thus the bilinear quantity

$$W(x,\xi) = \bar{\bar{\psi}}(x,\xi)\,\psi(x,\xi) \tag{4.33}$$

is a scalar function defined on the bundle space $T^R(V_4)$, but it is not a true function on the base space V_4 of the bundle. Under Lorentz gauge transformations $W(x,\xi)$ transforms as a scalar function, however, in addition $W(x,\xi)$ possesses de Sitter gauge transformations properties expressed by

$$W'(x,\xi') = W(x,\xi) \tag{4.34}$$

which shows that the bilinear quantity (4.33) defined on $T^R(V_4)$ contains more - although gauge dependent - information than a function W(x) defined on space-time would possess. This observation may be taken as an indication that the fiber bundle formalism, based on the noncompact nonabelian gauge group SO(4,1) and the elementary length parameter R which it introduces into the geometry, contains elements for a description of extended hadrons which are more general than the probability density functions of conventional quantum mechanics describing point-like objects (electrons) in space-time [x)].

IV.2 De Sitter Gauge Formulation of Strong Interactions

In order to focus the attention on the strong interaction effects associated with fields defined on the de Sitter bundle we choose the base space of $T^R(V_4)$ to be flat Minkowski space-time M_4. We, furthermore, neglect effects arising from electromagnetic interactions in this section and return to a $U(1) \otimes USp(2,2)$ gauge invariant description for the combined electromagnetic and strong interactions in the context of a Lagrangian formulation presented in the next section.

A Cartan connexion on the bundle $T^R(M_4)$ associated with the de Sitter frame bundle $L^R(M_4)$ over Minkowski space is, in analogy to eqs. (3.99) and (3.100) and in view of eq. (4.26), defined by an $\mathfrak{so}(4,1)$-valued one-form, ω^R , on M_4 with matrix elements

$$\omega^R_{ab}(x) = -\omega^R_{ba}(x) \quad ; a,b = 0,1,2,3,5 \, . \tag{4.35}$$

The $\omega^R_{ab}(x)$ determine, in a chosen gauge, the de Sitter rotation of the local frames on $T^R(M_4)$ in going from a point $x \in M_4$ to an infinitesimally neighboring point according to the formula

$$d\, e_a(\xi) = \omega^{R\ b}_{a}(x)\, e_b(\xi) \tag{4.36}$$

where on the left-hand side the symbol d does not denote the exterior derivative since the $e_a(\xi)$ are no forms, and

$$\omega^{R\ b}_{a}(x) = \omega^R_{ac}(x)\, \eta^{cb} \, . \tag{4.37}$$

---

x) Compare ref. 34 in this context.

Expanding the one-forms (4.35) in a basis of one-forms in Minkowski space provided by the coordinate differentials dx^μ one obtains

$$\omega^R_{ab}(x) = dx^\mu \, \Gamma^R_{\mu ab}(x) \tag{4.38}$$

where

$$\Gamma^R_{\mu ab}(x) = - \Gamma^R_{\mu ba}(x) \tag{4.39}$$

represent the fourty coefficients of a Cartan connexion on the bundle $T^R(M_4)$ which we shall call the de Sitter rotation coefficients. Taking the exterior derivative of (4.36) and calling

$$d\,(\, d\, e_a(\xi)) = \Omega^{R\ b}_{a}(x)\, e_b(\xi) \tag{4.40}$$

with $\Omega^{R\ b}_{a}$ being the matrix of the de Sitter curvature two-forms on $T^R(M_4)$, one at once establishes Cartan's structural equation for the curvature two-forms $\Omega^R_{ab} = \Omega^{R\ c}_{a}\,\eta_{cb}$, i.e.

$$d\omega^R_{ab} + \omega^R_{ac} \wedge \omega^{R\ c}_{b} = \Omega^R_{ab} = - \Omega^R_{ba} \; . \tag{4.41}$$

These equations correspond to eqs. (3.134) in the de Sitter case.

To define a de Sitter gauge invariant differentiation process for a spinor quantity defined on $T^R(M_4)$ we turn now to the spinor connexion, $\Gamma^R(x)$, being the connexion on the spinor bundle S^R of eq. (4.28) associated to the bundle of frames $L^R(M_4)$. $\Gamma^R(x)$ is, in a chosen gauge, a matrix valued one-form on M_4 with values in the Lie algebra of the group USp(2,2). In the basic 4x4 representation adapted to the four component de Sitter spinor fields (4.29) or (4.31) one has

$$\Gamma^R(x) = \frac{1}{2}\, \omega^R_{ab}(x)\, M^{ab} \tag{4.42}$$

with M^{ab} as given by eq. (4.27). Expanding this as

$$\Gamma^R(x) = dx^\mu \, \Gamma^R_\mu(x) \tag{4.43}$$

one obtains the following matrix valued four-vector coefficients, $\Gamma^R_\mu(x)$, involving the de Sitter rotation coefficients introduced in eq. (4.38) above

$$\Gamma_\mu^R(x) = \frac{1}{2}\Gamma_{\mu ab}^R(x)\, M^{ab} . \tag{4.44}$$

One can now multiply eq. (4.41) on both sides by M^{ab} and sum over a and b to give these equations, in view of the commutation relations of SO(4,1) expressed by eq. (4.8), the following matrix form

$$\mathcal{R}^R_{\mu\nu}(x) = \partial_\mu \Gamma_\nu^R(x) - \partial_\nu \Gamma_\mu^R(x) + i\left[\Gamma_\mu^R(x), \Gamma_\nu^R(x)\right] . \tag{4.45}$$

Here we have put

$$\frac{1}{2} dx^\mu \wedge dx^\nu \mathcal{R}^R_{\mu\nu}(x) = \frac{1}{4} dx^\mu \wedge dx^\nu R^R_{\mu\nu ab}(x) M^{ab} = \frac{1}{2}\Omega^R_{ab}(x) M^{ab} \tag{4.46}$$

where $R^R_{\mu\nu ab}(x)$ is the de Sitter curvature tensor defined on $T^R(M_4)$ which, in terms of the $\Gamma^R_{\mu ab}(x)$, can be expressed as

$$R^R_{\mu\nu ab} = \partial_\mu \Gamma^R_{\nu ab} - \partial_\nu \Gamma^R_{\mu ab} + \Gamma^R_{\mu ac}\Gamma^{R\;c}_{\nu b} - \Gamma^R_{\nu ac}\Gamma^{R\;c}_{\mu b} , \tag{4.47}$$

with $\Gamma^{R\;c}_{\mu a} = \Gamma^R_{\mu ab}\,\eta^{bc}$, which represent the third version of eqs. (4.41). From eqs. (4.39) and (4.47) one reads off the following symmetry relations for the tensor $R^R_{\mu\nu ab}(x)$:

$$R^R_{\mu\nu ab}(x) = -R^R_{\nu\mu ab}(x) = -R^R_{\mu\nu ba}(x) . \tag{4.48}$$

We finally quote the Bianchi identities for the curvature tensor (4.47) which are easily deduced from the exterior derivative of eq. (4.41) leading to $D\Omega^R_{ab}=0$ (compare eq. (3.86)) resulting in the equations

$$D_\varkappa R^R_{\mu\nu ab}(x) + D_\mu R^R_{\nu\varkappa ab}(x) + D_\nu R^R_{\varkappa\mu ab}(x) = 0 \tag{4.49}$$

with [x)]

---

x) In the matrix form (4.45) the covariant derivative of the de Sitter curvature tensor is expressed as

$$D_\varkappa \mathcal{R}^R_{\mu\nu} = \partial_\varkappa \mathcal{R}^R_{\mu\nu} + i\left[\Gamma^R_\varkappa, \mathcal{R}^R_{\mu\nu}\right] .$$

$$D_{\varkappa} R^{R}_{\mu\nu ab} = \partial_{\varkappa} R^{R}_{\mu\nu ab} - \Gamma^{R\,c}_{\varkappa a} R^{R}_{\mu\nu cb} - \Gamma^{R\,c}_{\varkappa b} R^{R}_{\mu\nu ac} \; . \tag{4.50}$$

Under de Sitter gauge transformations $\hat{S}(x)$ (see eqs. (4.30)) the spinor connexion transforms as

$$\Gamma'^{R}(x) = \hat{S}(x)\,\Gamma^{R}(x)\,\hat{S}^{-1}(x) - i\,\hat{S}(x)\,d\hat{S}^{-1}(x) \tag{4.51}$$

which is a formula analogous to eqs. (3.54') and (3.110). The curvature tensor (4.45) transforms homogenously under de Sitter gauge transformations, i.e.

$$\mathcal{R}'^{R}_{\mu\nu}(x) = \hat{S}(x)\,\mathcal{R}^{R}_{\mu\nu}(x)\,\hat{S}^{-1}(x) \quad . \tag{4.52}$$

The spinor connexion (4.42) enables one to write down the following de Sitter gauge invariant differentiation operation for a spinor quantity $\Psi(x,\xi)$ defined on the bundle space [x)]

$$D\Psi(x,\xi) = \left(d + i\,\Gamma^{R}(x)\right)\Psi(x,\xi) \tag{4.53}$$

with the matrix valued form $\Gamma^{R}(x)$ acting on the primed (i.e. de Sitter) spinor index of $\Psi(x,\xi)$. We mention in passing that for the base space of the bundle being a curved space-time manifold V_4 eq. (4.53) would in addition contain the spinor connexion $\Gamma(x)$, defined in eq. (3.101), acting on the unprimed (i.e. Dirac) spinor index of the bispinor quantity $\Psi(x,\xi)$. Including also electromagnetic effects the complete covariant derivative of $\Psi(x,\xi)$ representing charged bispinor objects on $T^{R}(V_4)$ would read

$$D\Psi(x,\xi) = \left(d + ieA(x) + i\,\Gamma(x) + i\,\Gamma^{R}(x)\right)\Psi(x,\xi) \tag{4.54}$$

with $A(x)=dx^{\mu}A_{\mu}(x)$ (compare also Sect. IV.3 for the inclusion of electromagnetic

---

[x)] We simply write $\Psi(x,\xi)$ for $\Psi(x, e_i ; \xi, e_a(\xi))$ defined in eq. (4.31) where in Minkowski space e_i ; i=0,1,2,3, represents a fixed reference frame.

effects). After this digression we return to eq. (4.53) and write down the commutator of two de Sitter gauge invariant derivatives applied to $\Psi(x,\xi)$, i.e.

$$[\mathcal{D}_\mu, \mathcal{D}_\nu]\,\Psi(x,\xi) = i\,\mathcal{R}^R_{\mu\nu}(x)\,\Psi(x,\xi) \tag{4.55}$$

which is identical to eq. (2.65) obtained in Sect. II.2.

With eqs. (4.43) and (4.53) it is now a simple matter to write down a Lorentz invariant and de Sitter gauge invariant equation of Dirac type for the bispinor defined on $T^R(M_4)$ x)

$$\gamma^\mu\left[\partial_\mu + i\,\Gamma^R_\mu(x)\right]\Psi(x,\xi) = -i m\,\Psi(x,\xi) \tag{4.56}$$

where m denotes the mass of the particle described by $\Psi(x,\xi)$. We mention in passing that there is still another invariant equation which could be called in this context the equation of motion in the fiber. Dirac [35] wrote it down in 1935. It was intended as an equation for electrons in a de Sitter model for the universe. Formulating it for our bispinor field $\Psi(x,\xi)$ this equation reads

$$\frac{1}{2R}\,\gamma^a\gamma^b\,L_{ab}(\xi)\,\Psi(x,\xi) = \left(\mu + \frac{2i}{R}\right)\Psi(x,\xi) \tag{4.57}$$

with xx)

$$L_{ab}(\xi) = i\,(\xi_a\partial_b - \xi_b\partial_a) \tag{4.58}$$

being a set of ten differential operators, satisfying eqs. (4.8), turning functions defined on V_4' again into functions defined on V_4' (see eq. (4.2)). The surprising property of eq. (4.57) is the appearance of an imaginary contribution of order $\frac{1}{R}$ to the mass μ which is a further constant with the dimension of an inverse length contained in the de Sitter bundle formalism if eq. (4.57) is demanded as equation of motion in the fiber. One could, however, also try to formulate the theory by putting the mass μ in eq. (4.57) equal to zero such that the scale of length is determined by R alone. In our further discussion in this paper eq. (4.57) will not play a prominent role. Thus we will not decide this question here.

---

x) To obtain the adjoint equation one uses $\gamma^0(i\Gamma^R_{(x)})^\dagger\gamma^0 = -i\,\Gamma^R_{(x)}$ implied by eq. (4.22), or, expressed for the generators, $M^{ab\dagger} = \gamma^0 M^{ab}\gamma^0$ as implied by eq. (4.19).

xx) $\partial_a = \frac{\partial}{\partial\xi^a}$.

To be able to solve eq. (4.56) the connexion coefficients $\Gamma^{R}_{\mu ab}(x)$, representing a set of fourty gauge potentials characterizing the geometry in the bundle space, have to be known as functions of x. They appear in eq. (4.56) so to speak as external fields determined by the geometry. In order to formulate a dynamics for strong interactions it is now suggestive to introduce the idea that the detailed behaviour and the values for $\Gamma^{R}_{\mu ab}(x)$ are to be determined from the matter function $\Psi(x,\xi)$ itself, i.e. to assume that the $\Gamma_{\mu}^{R}(x)$ of eq. (4.44) are quantities induced locally in the underlying fiber bundle geometry by the presence of matter in the geometry represented by $\Psi(x,\xi)$. This is our dualism, described in the introduction, of matter influencing the underlying geometry with this geometry playing an active part in the dynamics not merely providing an arena for the physics contained in the interaction of certain fields defined on a pregiven space-time. We thus want to introduce here an interaction of matter fields $\Psi(x,\xi)$ and $\bar{\Psi}(x,\xi)$ with degrees of freedom characterizing the underlying bundle geometry in which these fields are immersed and call the dynamics resulting therefrom the strong fiber dynamics (SFD). To obtain a possible mechanism for such an induction phenomenon of geometric fields depending on the detailed form and strength of a matter distribution present at a point x of space-time we compare the situation in strong interaction physics with the familiar example of electromagnetic fields interacting with charged matter, or that of the gravitational field interacting with gravitating matter. In a similar way as, on the one hand, in Maxwell's theory (compare eqs. (2.13)) where the electromagnetic fields $F_{\mu\nu}(x)$ are considered as generated by an electromagnetic current distribution, $j_{\mu}(x)$, of charged matter given in the Dirac theory by eq. (2.14), and, on the other hand, in general relativity where through Einstein's equations x) the Riemannian curvature field of space-time is coupled to the energy momentum distribution, $T_{\mu\nu}(x)$, of gravitating matter, we would like to regard here for a description of strong interactions the internal or de Sitter curvature tensor $R^{R}_{\mu\nu ab}(x)$, defined by eq. (4.47), as generated locally by a source current derived from the matter field $\Psi(x,\xi)$ and its adjoint. Integrating a density constructed from the bilocal quantities $\Psi(x,\xi)$ and $\bar{\Psi}(x,\xi)$ over the local fiber $F_x = V_4'(x)$ to obtain a formally local current expression we suggest the following quantities to act as a source current for the de Sitter curvature field on the bundle space

---

x) $R_{\mu\nu}(x) - \frac{1}{2} g_{\mu\nu}(x) R(x) = \varkappa T_{\mu\nu}(x)$,

where $R_{\mu\nu}(x) = g^{\varkappa\lambda}(x) R_{\varkappa\mu\lambda\nu}(x)$ denotes the Ricci tensor and $R(x) = g^{\mu\nu}(x) R_{\mu\nu}(x)$ the curvature scalar.

$$\mathcal{J}_{\mu ab}(x) = \int_{V_4'(x)} \bar{\bar{\Psi}}(x,\xi)\,(\gamma_\mu \otimes M_{ab})\,\Psi(x,\xi)\,d\mu(\xi). \qquad (4.59)$$

Here $d\mu(\xi) = \frac{R}{|\xi^5|}\, d\xi^0 \wedge d\xi^1 \wedge d\xi^2 \wedge d\xi^3$ is the invariant measure on the coset space SO(4,1)/SO(3,1) isomorphic to $V_4'(x)$. Since $V_4'(x)$ is compact in its spacial extensions and noncompact in "time" (i.e. in the ξ^0-direction) the support properties of $\Psi(x,\xi)$ on $V_4'(x)$ are required to be such that the integral (4.59) converges. We assume this to be case x). From eqs. (4.17) and (4.30) one establishes at once the following tensor transformation rule for the $\mathcal{J}_{\mu ab}(x)$ under de Sitter gauge transformations:

$$\mathcal{J}'_{\mu a'b'}(x) = [A^{-1}(x)]^{a}_{a'}\,[A^{-1}(x)]^{b}_{b'}\,\mathcal{J}_{\mu ab}(x). \qquad (4.60)$$

Furthermore, one has xx)

$$\mathcal{J}^{\dagger}_{\mu ab}(x) = \mathcal{J}_{\mu ab}(x) = -\mathcal{J}_{\mu ba}(x). \qquad (4.61)$$

$\mathcal{J}_{\mu ab}(x)$ are thus the components of an antisymmetrical, hermitean, tensor gauge current given in the local embedding space $R_{4,1}(x)$ and associated with the hypersurface $V_4'(x)$ transforming as a four-vector under Lorentz transformations. Under a de Sitter gauge transformation the indices a,b of $\mathcal{J}_{\mu ab}(x)$ undergo a local de Sitter rotation, i.e. a hyperbolic rotation in the local $R_{4,1}(x)$. It follows from the Dirac-type equation (4.56) and its adjoint together with the commutation rules (4.8) for the M^{ab} that the $\mathcal{J}_{\mu ab}(x)$ are covariantly conserved i.e. obey

$$D^{\mu}\,\mathcal{J}_{\mu ab}(x) = 0, \qquad (4.62)$$

---

x) For a more detailed discussion of the integral geometry in spaces of constant curvature see the book by I.M. Gelfand, M.I. Graev and N.Ya. Vielenkin(36).

xx) To compute the adjoint of (4.59) we regard here $\Psi(x,\xi)$ as a generalized wave function. For the operator character of the currents $\mathcal{J}_{\mu ab}(x)$ see the discussion given below.

where the covariant divergence is given in analogy to eq. (4.50) by (suppressing again the argument (x))

$$D^{\mu} \mathcal{J}_{\mu ab} = \partial^{\mu} \mathcal{J}_{\mu ab} - \Gamma^{R\mu}{}_{a}{}^{c} \mathcal{J}_{\mu cb} - \Gamma^{R\mu}{}_{b}{}^{c} \mathcal{J}_{\mu ac} \tag{4.63}$$

with $\Gamma^{R\mu}{}_{a}{}^{c} = \eta^{\mu\nu} \Gamma^{R}_{\nu ab} \eta^{bc}$ and $\eta^{\mu\nu}$ = diag (1,-1,-1,-1), the latter being the metric tensor in M_4.

We now propose the gauge dependent conserved local current (4.59) to act as a source current for the internal curvature field on the bundle space, i.e. we propose as dynamical equations representing the described dualism between hadronic matter fields and the underlying fiber geometry the following gauge invariant equations

$$D^{\mu} R^{R}_{\mu\nu ab}(x) = \bar{\varkappa}\, \mathcal{J}_{\nu ab}(x) \quad . \tag{4.64}$$

These are our main equations defining mathematically what we named strong fiber dynamics. For brevity we call these equations the current-curvature equations (c.-c. equations)[x)]. They are partial differential equations of second order for the strong interaction gauge potentials $\Gamma^{R}_{\mu ab}(x)$. It is easy to show that the conservation laws (4.62) follow directly from (4.64) due to the antisymmetry of $R^{R}_{\mu\nu ab}(x)$ in μ and ν .

The question now immediately poses itself whether the set of strong interaction field equations consisting of eqs. (4.64), the associated Bianchi identieis (4.49), and the Dirac-type equation (4.56), together with eqs. (4.47) relating the gauge potentials appearing in (4.56) to the curvature tensor appearing in the c.-c. equations are, in fact, unique. One could think of constructing another source current obeying analogous relations as do the $\mathcal{J}_{\mu ab}(x)$ of eq. (4.59) by using in the construction instead of the operators M_{ab} the orbital operators $L_{ab}(\xi)$ defined in eq. (4.58). It turns out, however, that such orbital-type source currents are forbidden to occur by time reversal invariance in Minkowski space required to hold for the strong interactions formulated here geometrically on the de Sitter-Cartan bundle. Thus, while eq. (4.64) is T-invariant, a similar equation with an L-type current on the right-hand side would not be T-invariant. For details of the argument we refer to the discussion given in ref. 32. The reason essentially is that in the spinor connexion matrices $\Gamma^{R}_{\mu}(x)$ only the generators M^{ab} appear without their orbital counterparts $L_{ab}(\xi)$.

---

x) We shall see below that the coupling constant $\bar{\varkappa}$ has the length dimension $[\ell^{-3}]$ (ℓ = length). In Sect. IV.3 we shall, moreover, introduce a dimensionless coupling constant g, to be related to strong interactions, by $\bar{\varkappa} = g/R^3$.

There is one further discrete symmetry which has to be satisfied by the proposed current-current equations. Up to now we only discussed hadronic matter represented by a bispinor field $\Psi(x,\xi)$ defined on the de Sitter bundle space. However, if there is a matter field $\Psi(x,\xi)$ affecting the underlying geometry of the bundle space through a source current in a way expressed by eq. (4.64) it is a legitimate question to ask how the geometry is influenced by antimatter the existence of which is an experimental fact which must be taken into account in the formalism by introducing a separate bispinor function $\Psi^{\hat{C}}(x,\xi)$ for it. Since the strong interactions are usually assumed to be the same among particles as they are among antiparticles we demand in the geometric formalism used here that the discrete symmetry $\hat{C}$ relating $\Psi^{\hat{C}}(x,\xi)$ and $\Psi(x,\xi)$ — called the generalized charge conjugation, or better, matter-antimatter conjugation — should leave the basic c.-c. equations (4.64) invariant. To discuss the discrete symmetry $\hat{C}$ for bispinor quantities defined on the bundle space we first factorize the representation character of $\Psi(x,\xi)$ with respect to the Lorentz group, operating in the base space, and the de Sitter gauge group, operating in the fiber, according to

$$\Psi(x,\xi) = \varphi(x)\, \phi_x(\xi) \quad . \tag{4.65}$$

Here $\varphi(x)$ represents the Dirac spinor part of $\Psi(x,\xi)$ being the conventional local Dirac spinor field associated with point-like particles like electrons in quantum electrodynamics . At first $\varphi(x)$ can be introduced here as a Dirac wave function in Minkowski space (c-number), however, then the imposition of $\hat{C}$-conjugation invariance of eqs. (4.64) (see below) forces one to regard $\varphi(x)$ as an operator (q-number) [x)]. In the absence of other than strong interactions we require $\varphi(x)$ to obey the free Dirac equation

$$\gamma^{\mu}\partial_{\mu}\varphi(x) + i m \varphi(x) = 0 \quad . \tag{4.66}$$

The second factor in eq. (4.65) represents the internal or de Sitter part of the total wave field $\Psi(x,\xi)$, with $\phi_x(\xi)$ being a short hand notation for $\phi(x;\xi,e_a(\xi))$ defined in eq. (4.29). The essential point is that the de Sitter part $\phi_x(\xi)$ in the factorization (4.65), coupling to the gauge potentials $\Gamma^{R}_{\mu ab}(x)$, is still x-dependent. The global behaviour of $\phi_x(\xi)$ on the de Sitter bundle space is expected to depend on the property of $T^R(M_4)$ (or $T^R(V_4)$ for this matter) being a trivial or a nontrivial fiber bundle over space-time. This is a difficult and fundamental question to answer. We know of no better way in trying to solve this problem than

---

[x)] Compare the more detailed discussion given in ref. 32.

to formulate strong interaction physics in every detail in the porposed geometric language and see whether the solutions of the established nonlinear equations enable one to make a decision. At this point of our discussion we, however, prefer to leave the question of the triviality or nontriviality of the underlying bundle open until explicit solutions are known.

We shall assume in the following that $\varphi(x)$ possesses the canonical length dimension $[\ell^{-3/2}]$ while $\phi_x(\xi)$ has subcanonical dimension $[\ell^{-1/2}]$ such that the field $\psi(x,\xi)$ defined on the de Sitter bundle carries the dimension $[\ell^{-2}]$ x).

The discrete symmetry transformation $\hat{C}$ leaving eqs. (4.64) invariant can now be seen to be an ordinary charge conjugation C for the space-time part $\varphi(x)$ taken together with an "internal" $\hat{P}\hat{T}$-transformation for the fiber, i.e. de Sitter, part $\phi_x(\xi)$ of $\psi(x,\xi)$ [32]. We thus write the $\hat{C}$-transformation for $\psi(x,\xi)$, with $\psi(x,\xi)$ given in the factorized form (4.65), as xx)

$$\psi^{\hat{C}}(x,\hat{I}_{PT}\xi) = \varphi^{C}(x)\,\phi_x^{\hat{P}\hat{T}}(\hat{I}_{PT}\xi) = \hat{\eta}_{PT}\left(C \otimes i\gamma^5\right)\tilde{\bar{\varphi}}(x)\,\phi_x(\xi) \tag{4.67a}$$

$$\bar{\bar{\psi}}^{\hat{C}}(x,\hat{I}_{PT}) = \bar{\varphi}^{C}(x)\,\bar{\phi}_x^{\hat{P}\hat{T}}(\hat{I}_{PT}\xi) = \hat{\eta}^{*}_{PT}\,\tilde{\varphi}(x)\,\bar{\phi}_x(\xi)\left(C^{-1}\otimes i\gamma^5\right) \tag{4.67b}$$

with C being the charge conjugation matrix of the conventional Dirac theory obeying

$$C\,\gamma^\mu\, C^{-1} = -\tilde{\gamma}^\mu \tag{4.68}$$

and

$$C = -C^{-1} = -C^{\dagger} = -\tilde{C} \tag{4.69}$$

where $\sim$ denotes the transposed matrix. $\hat{\eta}_{PT}$, with $\hat{\eta}_{PT} = \pm 1$ or $\pm i$ is the $\hat{C}$-conjugation parity, and

x) As mentioned before this leads to the length dimension $[\ell^{-3}]$ for the coupling constant $\varkappa$ in eqs. (4.64) since $\mathfrak{g}_{\mu ab}(x)$, as defined by eq. (4.59), is dimensionless and $\Gamma_\mu^R(x)$ has dimension $[\ell^{-1}]$ as is apparent from eq. (4.53). Correspondingly, the $\mathcal{R}^R_{\mu\nu}(x)$, defined by eqs. (4.45) possess the dimension $[\ell^{-2}]$.

xx) Again the direct product notations is used here for the Dirac and de Sitter matrices with the first factor associated with $\varphi(x)$ and the second factor associated with $\phi_x(\xi)$.

$$\hat{I}_{PT} = \begin{pmatrix} -1 & & & & \\ & -1 & & & \\ & & -1 & & \\ & & & -1 & \\ & & & & 1 \end{pmatrix} \tag{4.70}$$

is the "internal" $\hat{\hat{P}}\hat{\hat{T}}$-transformation matrix represented in terms of four-component spinors (without a doubling of the representation character being necessary) by the matrix $i\gamma^5$.

An essential result following from the demanded $\hat{C}$-invariance of eqs. (4.64) is the fact that the Dirac point spinor part $\varphi(x)$ of $\Psi(x,\xi)$ is required to possess q-number properties (i.e. anticommutation character), and that $\hat{C}$ is a combined symmetry on the bundle space composed of the discrete symmetry C in the base space of our bundle taken together with the discrete symmetry $\hat{\hat{P}}\hat{\hat{T}}$ in all the "internal" de Sitter spaces, i.e. the fibers, of the bundle. We thus, finally, write the current (4.59), using (4.65), in the following antisymmetrized form with respect to $\varphi(x)$ and $\bar{\varphi}(x)$:

$$\mathcal{J}_{\mu ab}(x) = \frac{1}{2}\left[\bar{\varphi}(x)\gamma_\mu\varphi(x) - \tilde{\varphi}(x)\tilde{\gamma}_\mu\tilde{\bar{\varphi}}(x)\right]\int_{V_4'(x)} \bar{\phi}_x(\xi)\, M_{ab}\,\phi_x(\xi)\, d\mu(\xi). \tag{4.71}$$

In a condensed notation this can be written as

$$\mathcal{J}_{\mu ab}(x) = j_\mu^D(x)\; F_{ab}(x) \tag{4.72}$$

where $j_\mu^D(x)$ represents the canonical q-number Dirac part which can be regarded, when taken separately, as an operator current, odd under C, for Dirac point particles in Minkowski space, i.e.

$$\begin{aligned} j_\mu^D(x) &= \frac{1}{2}\left[\bar{\varphi}(x)\gamma_\mu\varphi(x) - \tilde{\varphi}(x)\tilde{\gamma}_\mu\tilde{\bar{\varphi}}(x)\right] = \\ &= \frac{1}{2}\left[\bar{\varphi}(x)\gamma_\mu\varphi(x) - \bar{\varphi}^C(x)\gamma_\mu\varphi^C(x)\right], \end{aligned} \tag{4.73}$$

and where $F_{ab}(x)$ is a c-number de Sitter part

$$F_{ab}(x) = \int_{V_4'(x)} \bar{\phi}_x(\xi)\, M_{ab}\,\phi_x(\xi)\, d\mu(\xi) \tag{4.74}$$

originating from the underlying fiber geometry. The $F_{ab}(x)$ entering the gauge dependent current operator (4.72) could be called the form factor functions. Because of eq. (4.62) and the conservation of the Dirac current (4.73), following from eqs.

(4.66) and its adjoint, they satisfy

$$\mathcal{D}_\mu F_{ab}(x) = \partial_\mu F_{ab}(x) - \Gamma^{R\ \ c}_{\mu a}(x) F_{cb}(x) - \Gamma^{R\ \ c}_{\mu b}(x) F_{ac}(x) = 0. \qquad (4.75)$$

We remarks in closing this section that the imposition of $\hat{C}$-invairance for the basic current-curvature equations has led us from a purely geometrically motivated wave function type of description for hadronic matter to a many particle operator description — at least with respect to the Dirac space-time part of the bispinor field $\Psi(x,\xi)$. However, the internal x) motion, considered in this bundle formalism as taking place in the fiber, still remains represented by c-number quantities giving rise to form factor functions with internal, i.e. de Sitter, gauge transformation properties. This c-number part of the currents (4.72) is thus seen to be more easily interpretable in a classical geometrical way although the freedom of choosing gauges imposes some restrictions on such a classical picture. At any rate, the fiber part of the currents $\mathcal{J}_{\mu ab}(x)$ as well as the fiber part of the generalized wave field $\Psi(x,\xi)$ does not seem to require a particle interpretation, at least not at this level. To what extend the c.-c. equations require q-number character for the gauge fields $\Gamma^R_{\mu ab}(x)$, being now coupled to quantized source fields, must be seen in solving the system of equations for baryonic states representing different numbers of Dirac core particles. This may lead possibly to a q-number content also for the de Sitter part $\phi_x(\xi)$ to which the connexion coefficients couple. As mentioned in the introduction, we take here an intermediate position and treat the fields induced in the underlying geometry at first as classical c-number fields. It has, of course, to be carefully investigated in the future at what level quantum effects for the gauge or geometrical fields become essential. The existence of mesons as quanta of some nuclear field — for example the $\Gamma^R_{\mu ab}(x)$ or the $R^R_{\mu\nu ab}(x)$ in the framework described here — makes it very likely that the quantum aspect of these fields, regarded here as fields induced in the geometry, is, indeed, a very essential aspect of their nature.

IV.3 U(1) ⊗ USp(2,2) Gauge Invariant Lagrange Theory

Knowing after the discussion given in Sects. IV.1 and IV.2 that it is interesting to investigate generalized spaces possessing a fiber bundle structure of Cartan type and to formulate a gauge theory for an interaction in particle physics on such a geometric structure, it is now attractive to reconsider the arising situation in a Lagrangian framework. The reason for doing this is not to define or establish the gauge theory in this way. Since a gauge theory is much more directly and clearly

---

x) We leave out the quotation marks.

conceived in using a geometric language. In fact, it was a geometrical motivation which has led us to consider a Cartan-type bundle over space-time as a possible candidate for a geometric stratum on which hadron physics could manifest itself in the way described. The notion of a connexion on the bundle supplying a collection of gauge fields and the concept of a curvature on the bundle space leading to the path dependence of quantities when transferred along a certain curve from a given point of the base space to another one are geometrically transparent concepts needing no additional support or justification from a Lagrangian formulation. However, the Lagrangian formalism adds some new aspects when the question of an energy density for a dynamical system describing extended hadrons in the proposed geometric manner is considered. It is for such a study that a Lagrangian formulation is indeed helpful.

To include also electromagnetic effects we shall investigate in this section a theory based on a Lagrangian density $\mathcal{L}'(x,\xi)$ invariant under the gauge group $U(1) \otimes USp(2,2)$ being the direct product of the gauge group U(1) of the electromagnetic interaction and the gauge group USp(2,2) for the strong interaction as described in Sects. IV.1 and IV.2, respectively. Following current usage one could speak of a unified gauge description of electromagnetic and strong interactions. However, such a terminology is misleading since the U(1) fiber and the de Sitter fiber (with the latter having the Cartan property, i.e. being tangent to the base space at each point of space-time) are two distinct local spaces and, corresponding to this, each factor in the direct product $U(1) \otimes USp(2,2)$ representing the total gauge group gives rise to a separate coupling constant: e for electromagnetism [U(1)] and g for strong interactions [USp(2,2)].

We start from the Lagrangian density $\mathcal{L}^{(0)}(x)$ for a second quantized free Dirac field denoted by $\varphi(x)$, being defined on Minkowski space-time M_4 (compare eq. (2.6) where we called the c-number Dirac wave function $\psi(x)$)

$$\mathcal{L}^{(0)}(x) = \frac{i}{2}\left[\bar{\varphi}(x)\gamma^{\mu}\overrightarrow{\partial}_{\mu}\varphi(x) - \bar{\varphi}(x)\overleftarrow{\partial}_{\mu}\gamma^{\mu}\varphi(x)\right] - m\,\bar{\varphi}(x)\varphi(x) \tag{4.76}$$

where $\overrightarrow{\partial}_{\mu}$ and $\overleftarrow{\partial}_{\mu}$ act to the right and left, respectively. The associated energy momentum tensor (without Belinfante-Rosenfeld symmetrization) is given by

$$T^{(0)}_{\mu\nu}(x) = \frac{\partial \mathcal{L}^{(0)}(x)}{\partial \partial^{\mu}\varphi(x)}\,\partial_{\nu}\varphi(x) = \frac{i}{2}\left[\bar{\varphi}(x)\gamma_{\mu}\overrightarrow{\partial}_{\nu}\varphi(x) - \bar{\varphi}(x)\overleftarrow{\partial}_{\nu}\gamma_{\mu}\varphi(x)\right] \tag{4.77}$$

leading to the energy density for a free Dirac field

$$T^{(0)}_{00}(x) = \frac{i}{2}\left[\varphi^{\dagger}(x)\overrightarrow{\partial}_{0}\varphi(x) - \varphi^{\dagger}(x)\overleftarrow{\partial}_{0}\varphi(x)\right]\,. \tag{4.78}$$

It is well-known that applying Fermi-Dirac quantization to the spinor field $\varphi(x)$ leads to a positive definite Hamiltonian operator $H^{(o)}_{p.p.}$ for a system of Dirac point particles (abbreviated by the subscripts p.p.) which can in terms of creation and annihilation operators of the field $\varphi(x)$ be expressed as

$$H^{(0)}_{p.p.} = \int T^{(0)}_{00}(x)\, d^3x = \sum_{s=\pm 1/2} \int \frac{d^3p}{2p_0} \left[a^{+}_s(p) a_s(p) + b^{+}_s(p) b_s(p)\right] p_0 \quad . \tag{4.79}$$

Here we have used the following Fourier expansions for $\varphi(x)$ and $\bar{\varphi}(x)$ x)

$$\varphi(x) = \frac{1}{(2\pi)^{3/2}} \sum_s \int \frac{d^3p}{2p_0} \left[a_s(p)\, u_s(p)\, e^{-ip_\mu x^\mu} + b^{+}_s(p)\, v_s(p)\, e^{ip_\mu x^\mu}\right] \tag{4.80a}$$

$$\bar{\varphi}(x) = \frac{1}{(2\pi)^{3/2}} \sum_s \int \frac{d^3p}{2p_0} \left[a^{+}_s(p)\, \bar{u}_s(p)\, e^{ip_\mu x^\mu} + b_s(p)\, \bar{v}_s(p)\, e^{-ip_\mu x^\mu}\right] \tag{4.80b}$$

with the Lorentz invariant anticommutation relations

$$\{a^{+}_s(p), a_{s'}(p')\} = \{b^{+}_s(p), b_{s'}(p')\} = 2p_0\, \delta^3(\vec{p}-\vec{p}\,')\, \delta_{ss'} \tag{4.81}$$

and all the other anticommutators being zero. In view of the anticommutation property of the fields $\varphi(x)$ and $\bar{\varphi}(x)$ one rewrites eq. (4.76), and analogously eq. (4.77), as

$$\mathcal{L}^{(0)}(x) = \frac{i}{2}\left\{\bar{\varphi}(x)\gamma^\mu \overrightarrow{\partial}_\mu \varphi(x) - \bar{\varphi}(x)\overleftarrow{\partial}_\mu \gamma^\mu \varphi(x)\right\}_a - m\, \bar{\varphi}(x)\varphi(x) \tag{4.82}$$

with the subscript a on the curly bracket denoting antisymmetrization in $\varphi(x)$ and $\bar{\varphi}(x)$ i.e.

$$\left\{\bar{\varphi}(x)\gamma^\mu \overrightarrow{\partial}_\mu \varphi(x)\right\}_a = \frac{1}{2}\left[\bar{\varphi}(x)\gamma^\mu \partial_\mu \varphi(x) - (\partial_\mu \tilde{\varphi}(x))\tilde{\gamma}^\mu \tilde{\bar{\varphi}}(x)\right] \tag{4.83}$$

and similarly for the second term xx).

The electromagnetic interaction is now introduced by the minimal coupling replacement (see eq. (2.9))

---

x) Our normalization for the u and v spinors is $\bar{u}_s(p)u_s(p) = -\bar{v}_s(p)v_s(p) = 2\,m$.

xx) The minimal replacement carried out in eq. (4.82) (see below) then automatically leads to a coupling of the electromagnetic potentials to the Dirac current (4.73) being odd under charge conjugation.

$$\overrightarrow{\partial}_\mu \longrightarrow \overrightarrow{\partial}_\mu + ieA_\mu(x) \tag{4.84a}$$

$$\overleftarrow{\partial}_\mu \longrightarrow \overleftarrow{\partial}_\mu - ieA_\mu(x) \tag{4.84b}$$

with e denoting the charge of the particle described by the field $\varphi(x)$ (i.e. $e=+|e|$ for positrons and $e=-|e|$ for electrons). Performing the replacements (4.84) in eq. (4.82) and adding the Lagrangian (2.10) of the free electromagnetic fields yields the U(1) gauge invariant Lagrangian of quantum electrodynamics (QED).

In view of the discussion in the first two sections of this chapter and the geometric motivation offered there in favour of a fiber bundle or gauge description of strong interactions we now perform instead of eqs. (4.84) the following replacements in eq. (4.82):

(A):

$$\varphi(x) \longrightarrow \Psi(x,\xi) = \varphi(x)\,\phi_x(\xi) \tag{4.85a}$$

$$\bar{\varphi}(x) \longrightarrow \bar{\bar{\Psi}}(x,\xi) = \bar{\varphi}(x)\,\bar{\phi}_x(\xi) \tag{4.85b}$$

(B):

$$\overrightarrow{\partial}_\mu \longrightarrow \overrightarrow{D}_\mu = \overrightarrow{\partial}_\mu + ieA_\mu(x) + i\,\Gamma^R_\mu(x) \tag{4.86a}$$

$$\overleftarrow{\partial}_\mu \longrightarrow \overleftarrow{D}_\mu = \overleftarrow{\partial}_\mu - ieA_\mu(x) - i\,\Gamma^R_\mu(x) \quad . \tag{4.86b}$$

Here $A_\mu(x)$ determines the U(1) part of the connexion related to $\varphi(x)$, and $\Gamma^R_\mu(x)$ determines the USp(2,2) part related to $\phi_x(\xi)$ [with the latter connexion possessing the Cartan property]. Both parts together represent the four-vector coefficients of a connexion on a spinor bundle possessing the total gauge or structural group $U(1) \otimes USp(2,2)$. The resulting $U(1) \otimes USp(2,2)$ gauge invariant interaction Lagrangian is given by

$$\begin{aligned} \mathcal{L}'(x,\xi) = \frac{i}{2}\Big\{ & \bar{\bar{\Psi}}(x,\xi)\gamma^\mu\big[\overrightarrow{\partial}_\mu + ieA_\mu(x) + i\,\Gamma^R_\mu(x)\big]\Psi(x,\xi) \\ & - \bar{\bar{\Psi}}(x,\xi)\big[\overleftarrow{\partial}_\mu - ieA_\mu(x) - i\,\Gamma^R_\mu(x)\big]\gamma^\mu\Psi(x,\xi)\Big\}_a \\ & - m\,\bar{\bar{\Psi}}(x,\xi)\Psi(x,\xi) \quad . \end{aligned} \tag{4.87}$$

Integrating $\mathcal{L}'(x,\xi)$ over the local fiber of the de Sitter subbundle yields, together with eq. (4.59), the expression

$$L'(x) = \int_{V_4'(x)} \mathcal{L}'(x,\xi)\, d\mu(\xi) = \int_{V_4'(x)} \mathcal{L}^{(0)}(x,\xi)\, d\mu(\xi) - \qquad (4.88)$$
$$- eA_\mu(x)\, j^\mu(x) - \tfrac{1}{2}\, \Gamma^R_{\mu ab}(x)\, \mathfrak{J}^{\mu ab}(x)$$

where $\mathcal{L}^{(0)}(x,\xi)$ is given by eq. (4.76) with $\varphi(x)$ and $\bar{\varphi}(x)$ replaced by $\Psi(x,\xi)$ and $\bar{\Psi}(x,\xi)$, respectively. The currents $j^\mu(x)$ and $\mathfrak{J}^{\mu ab}(x)$ are given by (raising and lowering greek or latin indices with $\eta_{\mu\nu}$ and η_{ab}, respectively)

$$j_\mu(x) = j^D_\mu(x)\; F(x) \qquad (4.89)$$

$$\mathfrak{J}_{\mu ab}(x) = j^D_\mu(x)\; F_{ab}(x) \qquad (4.72)$$

where $j^D_\mu(x)$ is the Dirac current (4.73), and F(x) and $F_{ab}(x)$ are structure functions, i.e. form factor functions, defined in terms of $\phi_x(\xi)$ and $\bar{\phi}_x(\xi)$ (with $F_{ab}(x)$ as introduced before) by

$$F(x) = \int_{V_4'(x)} \bar{\phi}_x(\xi)\, \phi_x(\xi)\, d\mu(\xi) \qquad (4.90)$$

$$F_{ab}(x) = \int_{V_4'(x)} \bar{\phi}_x(\xi)\, M_{ab}\, \phi_x(\xi)\, d\mu(\xi) \;. \qquad (4.74)$$

In analogy to spinor electrodynamics we now write down a full Lagrangian L(x) by adding to L'(x) the Lagrangians of the free U(1) and USp(2,2) curvature fields given by (compare eqs. (2.2) and (2.10))

$$\mathcal{L}(F) = -\tfrac{1}{4}\, F_{\mu\nu}(x)\, F^{\mu\nu}(x) \qquad (2.10)$$

and, analogously, (compare eqs. (4.45) and (4.47)) by

$$\mathcal{L}(\mathcal{R}) = -\tfrac{1}{4}\, \mathrm{Tr}\left[\mathcal{R}^R_{\mu\nu}(x)\, \mathcal{R}^{R\mu\nu}(x)\right] = -\tfrac{1}{8}\, R^R_{\mu\nu ab}(x)\, R^{R\mu\nu ab}(x). \qquad (4.91)$$

The following quantity possesses the length dimension $[\ell^{-1}]$

$$L(x) = L'(x) + R^3 \mathcal{L}(F) + \frac{R^3}{g} \mathcal{L}(\mathcal{R}) \quad . \tag{4.92}$$

Here we have multiplied the Lagrangian densities (2.10) and (4.91) by the volume R^3 in order to obtain an electromagnetic energy (in the case of $\mathcal{L}(F)$) and a strong interaction energy (in the case of $\mathcal{L}(\mathcal{R})$) which can be associated with a system of finite size particles. Moreover, we have multiplied $\mathcal{L}(\mathcal{R})$ in eq. (4.92) by a factor $\frac{1}{g}$ with g being a dimensionless coupling constant related to strong interactions. The reason that such a coupling constant appears here in connection with the "free" de Sitter curvature field is that, on the one hand, we did not (for geometrical reasons) write explicitely a coupling constant in front of $\Gamma_\mu^R(x)$ in eqs. (4.86); thus the constant absorbed in $\Gamma_\mu^R(x)$ is expected to occur elsewhere. That it, in fact, has to appear in front of the $\mathcal{L}(\mathcal{R})$ term in eq. (4.92) is, on the other hand, very reasonable from the point of view of the dualism operative in our formulation since the energy associated with the de Sitter curvature field $R^R_{\mu\nu ab}(x)$ is regarded as deposited locally in the fiber bundle geometry, i.e. near the point particle described by $\varphi(x)$. This situation is quite different from the corresponding situation in electrodynamics, or in the theory of gravitation, based on fields having an infinite range where it is, indeed, possible to consider the covariant field strengths far away from the sources. The de Sitter curvature field, on the other hand, represents the local response in the Cartan bundle geometry of a matter field present at x with the $R^R_{\mu\nu ab}(x)$ possessing a smooth distribution around x since a quantity defined on the bundle space cannot disappear discontinuously as a function of x. The resulting effect is determined on the level of the matter field by the de Sitter factor $\phi_x(\xi)$ contained in $\Psi(x,\xi)$, and on the level of the geometry by $R^R_{\mu\nu ab}(x)$. Thus, in view of these remarks and of what has been expressed in the introduction, the curvature fields $R^R_{\mu\nu ab}(x)$ cannot be considered as really free gauge field strengths. They are what we called induced fields which, so to speak, follow the point-like particle, described by $\varphi(x)$, like a shadow or halo in the geometry. Of course, the combined effect of the Dirac-de Sitter bispinor field $\Psi(x,\xi)$ together with the local curvature fields $R^R_{\mu\nu ab}(x)$ represents the physically observable (extended) hadron (compare again Fig. 1). Before we ask the question what the amount of energy actually is which is connected in such a picture with an isolated stable hadron we draw some conclusions from the expression for L(x).

Assuming that the equations of motion follow from the variational principle [x)]

---

x) Compare the discussion given in Chapter II. The Dirac-type equation (see below) would follow from the Lagrangian $\mathcal{L}'(x,\xi)$ (before integrating over the fibers) with the help of the variational principle

$$\delta \int_\Omega \mathcal{L}'(x,\xi)\, d^4x = 0.$$

$$\delta \int_{\Omega} L(x)\, d^4x = 0 \tag{4.93}$$

one finds by varying $A_\mu(x)$ and $\Gamma^R_{\mu ab}(x)$, with eqs. (4.92), (4.88), (4.89) and (4.72), the following current-curvature equations of electromagnetic and strong interaction type

$$\partial^\nu F_{\mu\nu}(x) = \frac{e}{R^3}\, j_\mu(x) = e\, j^D_\mu(x)\, \frac{1}{R^3} \int_{V'_4(x)} \bar{\Phi}_x(\xi)\, \phi_x(\xi)\, d\mu(\xi) \tag{4.94}$$

$$D^\nu R^R_{\nu\mu ab}(x) = \frac{g}{R^3}\, \mathcal{J}_{\mu ab}(x) = g\, j^D_\mu(x)\, \frac{1}{R^3} \int_{V'_4(x)} \bar{\Phi}_x(\xi)\, M_{ab}\, \phi_x(\xi)\, d\mu(\xi) \tag{4.95}$$

implying the conservation laws.

$$\partial^\mu j_\mu(x) = 0 \tag{4.96}$$

$$D^\mu \mathcal{J}_{\mu ab}(x) = 0 \;. \tag{4.62}$$

Furthermore, the following integrability conditions (Bianchi identities) must be satisfied

$$\partial_\varkappa F_{\mu\nu}(x) + \partial_\mu F_{\nu\varkappa}(x) + \partial_\nu F_{\varkappa\mu}(x) = 0 \tag{2.16}$$

$$D_\varkappa R^R_{\mu\nu ab}(x) + D_\mu R^R_{\nu\varkappa ab}(x) + D_\nu R^R_{\varkappa\mu ab}(x) = 0 \;. \tag{4.49}$$

Eqs. (4.94) and (2.16) represent the two groups of Maxwell equations in the present case for the electromagnetic fields being coupled to a source current, $j_\mu(x)$, describing extended particles. Eq. (4.96) expresses the conservation for the electric charge. The eqs. (4.95) and the Bianchi identities (4.49) are , on the other side, the current-curvature equations on the de Sitter bundle for our geometrically motivated theory of strong interactions with eq. (4.62) expressing the associated covariant conservation law which is more difficult to interprete than in the U(1) case due to the nonlinearity it involves.

We complete the list of equations characterizing the U(1) ⊗ USp(2,2) gauge invariant theory by writing down the Dirac-type equation following from eq. (4.66) derivable from $\mathcal{L}^{(0)}(x)$ (see eq. (4.76)) under the replacements (4.85) and (4.86):

$$\gamma^{\mu}\left[\partial_{\mu} + ieA_{\mu}(x) + i\,\Gamma_{\mu}^{R}(x)\right]\Psi(x,\xi) = -im\,\Psi(x,\xi) \quad . \tag{4.97}$$

So far we considered a matter field $\Psi(x,\xi)$ transforming as a spinor under internal, i.e. de Sitter, transformations. Hadrons - or, more exactly, baryons - were characterized by a de Sitter spin $j_{\text{de Sitter}} = \frac{1}{2}$. We mentioned briefly in the introduction what place leptons could find in a world having basically the geometry of a Cartan bundle. Representing leptons as de Sitter scalars, corresponding to $j_{\text{de Sitter}} = 0$, one at once concludes from the absence of the $\Gamma_{\mu}^{R}(x)$-term in the Dirac equation for leptons corresponding to eq. (4.97) and the vanishing of the right-hand side of eqs. (4.95) in this case that leptons do not, in fact, possess a strong fiber dynamics. Thus the dualistic interplay between matter fields and geometry characterizing hadrons is suspended for leptons being de Sitter scalars. On the other hand, the right-hand side of eq. (4.95) is nonzero even for internal scalars $\phi_x(\xi)$ (with adjoint $\phi_x^{*}(\xi)$) leading to an additional form factor effect in QED originating from the function F(x). However, this deviation from QED may be vanishingly small in a situation where the fiber dynamics, with $\phi_x(\xi)$ adjusting itself in response to an internal curvature and vice versa, is not operating.

Let us now, finally, turn to the energy problem to see what energy contribution the induced geometrical fields are able to make in the hadron case. The U(1) ⊗ USp(2,2) gauge invariant energy momentum tensor $T'_{\mu\nu}(x)$ of length dimension $[\ell^{-5}]$ associated with the matter field $\Psi(x,\xi)$ is, in analogy to eq. (4.77), given by

$$\begin{aligned} T'_{\mu\nu}(x,\xi) &= \frac{\partial \mathcal{L}'(x,\xi)}{\partial \mathcal{D}^{\mu}\Psi(x,\xi)}\,\mathcal{D}_{\nu}\Psi(x,\xi) = \\ &= \frac{i}{2}\left[\bar{\bar{\Psi}}(x,\xi)\gamma_{\mu}\overrightarrow{\mathcal{D}}_{\nu}\Psi(x,\xi) - \bar{\bar{\Psi}}(x,\xi)\overleftarrow{\mathcal{D}}_{\nu}\gamma_{\mu}\Psi(x,\xi)\right] . \end{aligned} \tag{4.98}$$

Integrating over the local fiber this can be written as (compare eq. (4.88))

$$\begin{aligned} T'_{\mu\nu}(x) &= \int_{V'_4(x)} T'_{\mu\nu}(x,\xi)\,d\mu(\xi) = \int_{V'_4(x)} T^{(0)}_{\mu\nu}(x,\xi)\,d\mu(\xi) - \\ &- e\,j_{\mu}(x)A_{\nu}(x) - \frac{1}{2}\,J_{\mu ab}(x)\,\Gamma_{\nu}^{R\,ab}(x) \end{aligned} \tag{4.99}$$

with $T^{(0)}_{\mu\nu}(x,\xi)$ as given by eq. (4.77) where $\varphi(x)$ and $\bar{\varphi}(x)$ being replaced by $\psi(x,\xi)$ and $\bar{\psi}(x,\xi)$, respectively. $T'_{\mu\nu}(x)$ is a quantity having the dimension of an inverse length associated with the matter field $\psi(x,\xi)$ interacting with the gauge potentials $A_\mu(x)$ and $\Gamma^R_{\mu ab}(x)$. To obtain the expression for the total energy eq. (4.99) has to be completed by adding the contribution of the gauge field strengths $F_{\mu\nu}(x)$ and $R^R_{\mu\nu ab}(x)$. In parallel to eq. (4.92) we now write down the following gauge invariant total energy quantity $T_{00}(x)$ associated with the system of fields $\psi(x,\xi)$, $A_\mu(x)$ and $\Gamma^R_{\mu ab}(x)$:

$$T_{00}(x) = T'_{00}(x) + R^3 T_{00}(F) + \frac{R^3}{g} T_{00}(\mathbb{R}) \tag{4.100}$$

where the energy densities $T_{00}(F)$ and $T_{00}(\mathbb{R})$ are given by (leaving out the argument x)

$$\begin{aligned} T_{00}(F) &= \left\{ \tfrac{1}{4} \eta_{\mu\nu} (F_{\varkappa\lambda} F^{\varkappa\lambda}) - F_{\mu\varkappa} F_\nu{}^{\varkappa} \right\}_{\substack{\mu=0\\ \nu=0}} \\ &= \tfrac{1}{2} [(\vec{E})^2 + (\vec{H})^2] , \end{aligned} \tag{4.101}$$

and

$$\begin{aligned} T_{00}(\mathbb{R}) &= \left\{ \tfrac{1}{8} \eta_{\mu\nu} (R^R_{\varkappa\lambda ab} R^{R\varkappa\lambda ab}) - \tfrac{1}{2} R^R_{\mu\varkappa ab} R^R_\nu{}^{\varkappa ab} \right\}_{\substack{\mu=0\\ \nu=0}} \\ &= \tfrac{1}{2} \Big\{ [(\vec{E}_{(m)})^2 + (\vec{H}_{(m)})^2] - \\ &\quad - [(\vec{E}_{(e)})^2 + (\vec{H}_{(e)})^2] - [(\vec{E}_{(b)})^2 + (\vec{H}_{(b)})^2] \Big\} . \end{aligned} \tag{4.102}$$

Here $\vec{E}$ and $\vec{H}$ denote, as usual, the electric and magnetic field strengths defined in Sect. II.1, and $E_{(q)}$ and $H_{(q)}$, with (q) = (e), (m), (b) standing for quasielectric (e), quasimagnetic (m) and boost (b) contributions contained in $R^R_{\mu\nu ab}(x)$ are defined by

$$E_{r,ab}(x) = R^R_{r0ab}(x) \quad ; \quad r = 1,2,3, \tag{4.103a}$$

$$H_{r,ab}(x) = R^R_{stab}(x) \quad ; \quad r,s,t \text{ cyclic} , \tag{4.103b}$$

with the a b-component called (e), (m) or (b) according to

$$
\begin{array}{llllll}
a,b = p,0 ; & p & = 1,2,3 & (e) & \\
a,b = p,q ; & p,q & = 1,2,3 \text{ cyclic} & (m) & \qquad (4.104)\\
a,b = 5,i ; & i & = 0,1,2,3 & (b) & .
\end{array}
$$

In total there are ten internal components for each field (4.103a) and (4.103b) labelled in the same way as the compact or noncompact generators of SO(4,1) shown in Fig. 5.

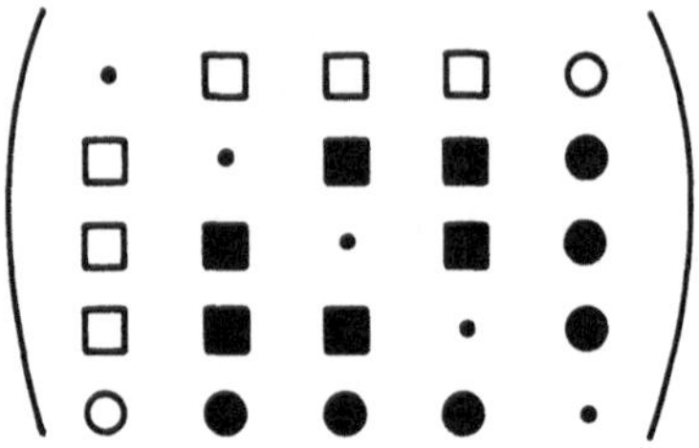

Compact (■, ●) and noncompact (□, ○) generators

$M_{ab} = -M_{ba}$ of SO(4,1)

M_{po}; $p = 1,2,3$	□	(e)
M_{pq}; $p \neq q$; $p,q = 1,2,3$	■	(m)
M_{50}	○	(b)
M_{5p}; $p = 1,2,3$	●	(b)

Fig. 5

The boost contribution in eq. (4.102) is, moreover, seen to give rise to two contributions i) a positive contribution for a,b = 5,p; p = 1,2,3 (denoted by $\bar{b}$ below), and ii) a negative contribution for a,b = 5,0 (denoted by b_o below). Thus one finds the following positive and negative contributions to the energy density $T_{oo}(\mathcal{R})$:

		Symbol in Fig. 5	Number of contributions	
positive	$\vec{E}_{(m)}, \vec{H}_{(m)}$	■	3	(4.105a)
	$\vec{E}_{(\bar{b})}, \vec{H}_{(\bar{b})}$	●	3	(4.105b)
negative	$\vec{E}_{(e)}, \vec{H}_{(e)}$	□	3	(4.105c)
	$\vec{E}_{(b_o)}, \vec{H}_{(b_o)}$	○	1	(4.105d)

From this it is apparent that the compact generators of the gauge group SO(4,1) give rise to six positive contributions to $T_{oo}(R)$ (and therefore to $T_{oo}(x)$) called to be of quasimagnetic and boost (three-vector part) type, and that the noncompact generators of SO(4,1) give rise to four negative contributions to $T_{oo}(R)$ (and hence to $T_{oo}(x)$) called to be of quasielectric and boost (time-part) type. It can be shown(37) by choosing a particular gauge that the first part on the right-hand side of eq.(4.100) gives rise to a positive contribution when computed for a one-particle state $|p,s\rangle = a_s^+(p)|0\rangle$ or a one-antiparticle state $|\overline{p,s}\rangle = b_s^+(p)|0\rangle$ of the field $\varphi(x)$ x). Also the second contribution on the right-hand side of eq. (4.100), given by eq. (4.101), is clearly positive. In view of eq. (4.105) the one-particle matrix elements of $T_{oo}(x)$ between states of equal momentum and spin projection, representing the energy associated with stable extended hadronic states, can thus in general only be positive definite if one restricts the noncompact strong interaction gauge group USp(2,2) to the compact subgroup SU(2) ⊗ SU(2) and, correspondingly, describes stable hadronic states of positive energy by representations of the maximal compact subgroup SU(2) ⊗ SU(2) of USp(2,2). The group SU(2) ⊗ SU(2) is the covering group of the subgroup SO(4) of SO(4,1) generated by the M_{ab} characterized by the symbols ■ and ● in Fig. 5.

A very interesting but at this point merely speculative question now is whether the noncompact generators of SO(4,1) can be associated with weak decays of hadrons changing possibly their representation character in the decay according to $\Delta j_{SO(4)} = 0$ or 1. Whether this can, indeed, be substanciated by detailed properties of the proposed geometrical model, in particular with regard to the weakness of these transitions, has still to be seen xx).

To conclude this review we can characterize our endeavour to establish a gauge formulation for extended hadrons using differential geometric techniques by the following remark. The geometrically motivated de Sitter bundle formalism, starting from the idea that the fiber of a bundle, characterized by an elementary length parameter R, is a space in which an internal motion of hadrons takes place and which possesses a tangent space isomorphic to Minkowski space such that the soldering of the fiber with space-time is possible (Cartan bundle) has, finally, led us to a SO(4) gauge symmetry for extended positive energy hadronic one-particle states.

---

x) $|0\rangle$ denotes the ground state in the Hilbert space of states, i.e. the state corresponding to an "empty geometry".

xx) The idea here is similar to the proposal to embed the SO(4) symmetry group of the nonrelativistic hydrogen atom into a bigger noncompact (so-called dynamical) group, for example SO(4,1). Compare refs. 38, 39 and the literature quoted there, and A. Böhm(40) for the introduction of such a concept into particle physics.

Acknowledgements

I am grateful to the University of Texas at Austin for the kind hospitality and for financial support and to the German Academic Exchange Service (DAAD) for a travel grant.

Bibliography

(1) S. Weinberg, Phys.Rev.Letters, 19, 1264 (1967);
A. Salam, in Elementary Particle Theory, edited by N. Svartholm, Almquist and Wiksell Förlag, Stockholm 1968;
E.S. Abers and B.W. Lee, Phys.Reports 9C, No 1 (1973).

(2) T.W.B. Kibble, J.Math.Phys. 2, 212 (1961);
D.W. Sciama, Recent Developments in General Relativity, Pergamon Press, London 1962, p. 415.
K. Hayashi and T. Nakano, Progr.Theor.Phys. 38, 491 (1967).

(3) C.N. Yang and R.L. Mills, Phys.Rev. 96, 191 (1954).

(4) R. Utiyama, Phys.Rev. 101, 1597 (1956).

(5) J.D. Bjorken and S.D. Drell, Relativistic Quantum Fields, Mac Graw Hill Book Company, New York 1965.

(6) S. Mandelstam, Ann. of Phys. 19, 1 (1962).

(7) T.T. Wu and C.N. Yang, Phys.Rev. D12, 3843 (1975).

(8) Y. Aharonov and D. Bohm, Phys.Rev. 115, 485 (1959).

(9) H. Weyl, Ber.d.preuss.Akad.d.Wissensch. 1918, p. 465, and Naturwiss. 19, 49 (1931).

(10) F. London, Naturwiss. 15, 187 (1927), and Z.f.Physik 42, 375 (1927).

(11) R.G. Chambers, Phys.Rev.Letters 5, 3 (1960).

(12) P.A.M. Dirac, Proc.Roy.Soc. (London) 133, 60 (1931).

(13) K. Przibram, Editor, Briefe zur Wellenmechanik, Springer Verlag Wien, 1963.

(14) S.W. Hawking and G.F.R. Ellis, Large Scale Structure of Space-Time, Cambridge University Press, 1973.

(15) K. Nomizu, Lie Groups and Differential Geometry, Publ.Math.Soc. Japan, 1956.

(16) S. Kobayashi and K. Nomizu, Foundations of Differential Geometry, Vol. 1, John Wiley, Interscience, New York, 1963.

(17) A. Lichnerowicz, Théorie globale des connexions et des groupes d'holonomie, Edizioni Cremonese, Roma 1962.

(18) Y. Choquet-Bruhat, Géometrie differentielle et systèmes extérieurs, Dunod, Paris 1968.

(19) S. Kobayashi, Theory of Connexions, Annali di Math. 43, 119 (1957).

(20) H. Weyl, Z.f.Physik, 56, 330 (1929).

(21) P.A. Carruthers, Spin and Isospin in Particle Physics, Gordon and Breach, New York 1973.

(22) R. Geroch, J.Math.Phys. 9, 1739 (1968).

(23) E. Cartan, Ann.Ec.Norm. 40, 325 (1922), Ann.Ec.Norm. 41, 1 (1924), and Leçons sur la géometrie des espaces de Riemann, Gauthier-Villars, Paris 1951.

(24) H. Flanders, Differential Forms, Academic Press, New York, 1963.

(25) L.P. Eisenhart, Riemannian Geometry, Princeton University Press, 1926.

(26) E. Cartan, Bull.Sc.Math.France 48, 294 (1924), Acta Math. 48, 1 (1926), Ann.of Math. 38, 1 (1937).

(27) Ch. Ehresmann, Colloque de Topologie (espaces fibrés), Bruxelles 1950, p. 29.

(28) R. Gilmore, Lie Groups, Lie Algebras, and Some of Their Applications, J. Wiley, New York 1974.

(29) S. Helgason, Differential Geometry and Symmetric Spaces, Academic Press, New York, 1962.

(30) W. Drechsler, Group Contraction in a Fiber Bundle with Cartan Connexion, J.Math.Phys. (to be published).

(31) W. Drechsler, Fortschr. der Physik, 23, 607 (1975).

(32) W. Drechsler, Currents in a Theory of Strong Interactions Based on a Fiber Bundle Geometry, Found.of Phys. (to be published).

(33) R. Takahashi, Bull.Soc.Math. France 91, 298 (1963).

(34) W. Drechsler, Is the Probability Concept still Fundamental in a Theory of Elementary Particles ? Phys.Letters 66B, 439 (1977).

(35) P.A.M. Dirac, Ann.Math. 36, 657 (1935).

(36) I.M. Gelfand, M.I. Graev and N.Ya. Vilenkin, Generalized Functions, Vol. 5, Chapter V, Academic Press, London 1966.

(37) W. Drechsler, Lagrange Formulation for a Gauge Theory of Strong and Electromagnetic Interactions Defined on a Cartan Bundle, Preprint MPI-PAE/PTh 45/76, December 1976.

(38) H. Bacry, Nuovo Cimento 41, 222 (1966).

(39) M. Bander and C. Itzykson, Rev.mod.Phys. 38, 330 (1966).

(40) A. Böhm, Phys.Rev. 145, 1212 (1966).

Lecture Notes in Physics

Vol. 44: R. A. Breuer, Gravitational Perturbation Theory and Synchrotron Radiation. VI, 196 pages. 1975.

Vol. 45: Dynamical Concepts on Scaling Violation and the New Resonances in e^+e^- Annihilation. Edited by B. Humpert. VII, 248 pages. 1976.

Vol. 46: E. J. Flaherty, Hermitian and Kählerian Geometry in Relativity. VIII, 365 pages. 1976.

Vol. 47: Padé Approximants Method and Its Applications to Mechanics. Edited by H. Cabannes. XV, 267 pages. 1976.

Vol. 48: Interplanetary Dust and Zodiacal Light. Proceedings 1975. Edited by H. Elsässer and H. Fechtig. XII, 496 pages. 1976.

Vol. 49: W. G. Harter and C. W. Patterson, A Unitary Calculus for Electronic Orbitals. XII, 144 pages. 1976.

Vol. 50: Group Theoretical Methods in Physics. 4th International Colloquium. Nijmegen 1975. Edited by A. Janner, T. Janssen, and M. Boon. XIII, 629 pages. 1976.

Vol. 51: W. Nörenberg und H. A. Weidenmüller. Introduction to the Theory of Heavy-Ion Collisions. IX, 273 pages. 1976.

Vol. 52: M. Mladjenović, Development of Magnetic β-Ray Spectroscopy. X, 282 pages. 1976.

Vol. 53: D. J. Simms and N. M. J. Woodhouse, Lectures on Geometric Quantization. V, 166 pages. 1976.

Vol. 54: Critical Phenomena. Sitges International School on Statistical Mechanics, June 1976. Edited by J. Brey and R. B. Jones. XI, 383 pages. 1976.

Vol. 55: Nuclear Optical Model Potential. Proceedings 1976. Edited by S. Boffi and G. Passatore. VI, 221 pages. 1976.

Vol. 56: Current Induced Reactions. International Summer Institute, Hamburg 1975. Edited by J. G. Körner, G. Kramer, and D. Schildknecht. V, 553 pages. 1976.

Vol. 57: Physics of Highly Excited States in Solids. Proceedings 1975. Edited by M. Ueta and Y. Nishina. IX, 391 pages. 1976.

Vol. 58: Computing Methods in Applied Sciences. Proceedings 1975. Edited by R. Glowinski and J. L. Lions. VIII, 593 pages. 1976.

Vol. 59: Proceedings of the Fifth International Conference on Numerical Methods in Fluid Dynamics. 1976. Edited by A. I. van de Vooren and P. J. Zandbergen. VII, 459 pages. 1976.

Vol. 60: C. Gruber, A. Hintermann, and D. Merlini, Group Analysis of Classical Lattice Systems. XIV, 326 pages. 1977.

Vol. 61: International School on Electro and Photonuclear Reactions I. Edited by C. Schaerf. VIII, 650 pages. 1977.

Vol. 62: International School on Electro and Photonuclear Reactions II. Edited by C. Schaerf. VIII, 301 pages. 1977.

Vol. 63: V. K. Dobrev et al., Harmonic Analysis on the n-Dimensional Lorentz Group and Its Application to Conformal Quantum Field Theory. X, 280 pages. 1977.

Vol. 64: Waves on Water of Vabriable Depth. Edited by D. G. Provis and R. Radok. 231 pages. 1977.

Vol. 65: Organic Conductors and Semiconductors. Proceedings 1976. Edited by L. Pál, G. Grüner, A. Jánossy and J. Sólyom. 654 pages. 1977.

Vol. 66: A. H. Völkel, Fields, Particles and Currents. VI, 354 pages. 1977.

Vol. 67: W. Drechsler and M. E. Mayer, Fiber Bundle Techniques in Gauge Theories. X, 248 pages. 1977.